Current Issues in Safety-Critical Systems

T0214172

Springer
London
Berlin
Heidelberg
New York
Hong Kong
Milan
Paris
Tokyo

Related titles:

Towards System Safety
Proceedings of the Seventh Safety-critical Systems Symposium, Huntingdon, UK 1999
Redmill and Anderson (Eds)
1-85233-064-3

Lessons in System Safety
Proceedings of the Eighth Safety-critical Systems Symposium, Southampton, UK 2000
Redmill and Anderson (Eds)
1-85233-249-2

Aspects of Safety Management
Proceedings of the Ninth Safety-critical Systems Symposium, Bristol, UK 2001
Redmill and Anderson (Eds)
1-85233-411-8

Components of System Safety
Proceedings of the Tenth Safety-critical Systems Symposium, Southampton, UK 2002
Redmill and Anderson (Eds)
1-85233-561-0

Felix Redmill and Tom Anderson (Eds)

Current Issues in Safety-Critical Systems

Proceedings of the Eleventh Safety-critical Systems Symposium, Bristol, UK 4-6 February 2003

Safety-Critical
Systems Club

Springer

Felix Redmill
Redmill Consultancy, 22 Onslow Gardens, London, N10 3JU

Tom Anderson
Centre for Software Reliability, University of Newcastle,
Newcastle upon Tyne, NE1 7RU

British Library Cataloguing in Publication Data
Safety-critical Systems Symposium (11th : 2003 : Bristol,
UK)
    Current issues in safety-critical systems : proceedings of
    the Eleventh Safety-critical Systems Symposium, Bristol, UK
    4-6 February 2003
    1.Industrial safety – Management – Congresses 2.Automatic
    control – Reliability – Congresses 3.Computer software –
    Reliability – Congresses
    I.Title II.Redmill, Felix, 1944- III.Anderson, T. (Thomas),
    1947-
    620.8'6
    ISBN-13:978-1-85233-696-7   e-ISBN-13:978-1-4471-0653-1
    DOI: 10.1007/978-1-4471-0653-1

Library of Congress Cataloging-in-Publication Data
A catalog record for this book is available from the Library of Congress.

ISBN-13:978-1-85233-696-7
a member of BertelsmannSpringer Science+Business Media GmbH
http://www.springer.co.uk

Typesetting: Camera ready by contributors

34/3830-543210 Printed on acid-free paper SPIN 10892815

PREFACE

There has been a great deal of development in the field of safety-critical systems since the first Safety-critical Systems Symposium was held in 1993. Our awareness of safety issues has been raised and our use of safety technologies has increased. Yet, every step in learning reveals the need for further learning and research, and opens doors onto both inadequacies and opportunities. This annual symposium offers the opportunity for problems to be exposed and discussed, solutions to be proposed and compared, and experiences to be reported, all to the benefit of the entire community.

In the last decade, there has been an increase in the extent to which safety practitioners have attempted to address the 'soft' issues, such as human factors, legal considerations, safety management, and safety assessment. The programme of SSS '93, and thus the content of this book, reflects this trend, and places a heavy emphasis on 'current issues in safety-critical systems'. The first paper reflects the symposium's tutorial on safety requirements, and the others are grouped into six sections:
- Human Factors
- The Safety Case - 1
- Development and Legal Issues
- The Safety Case - 2
- Safety Assessment
- Safety Standards

Of the 16 papers, 9 are from industry, 5 from academe, and 2 from research institutions. They present methods and ideas for safety improvement and reports on work in progress. Not only do their main themes offer lessons; often they draw attention to the safety implications of the small things that occurred, or were not in the first place thought of - the devil is in the detail. In addition to the 'practical' papers, one, by Armstrong and Paynter, takes a philosophical, linguistic, approach, and one, by Popat and Eastman, describes the government's proposals for legislation on corporate killing, something that potentially affects all of us.

We would like to thank the authors, both for writing their papers and for their cooperation in the preparation of the manuscript of this book. We also thank Joan Atkinson for her continued dedication to the work of the Safety-Critical Systems Club and for her efforts in ensuring the logistics of the symposium.

FR and TA
October 2002

THE SAFETY-CRITICAL SYSTEMS CLUB
sponsor and organiser
of the
Safety-critical Systems Symposium

What is the Club?
The Safety-Critical Systems Club exists to raise awareness and facilitate technology transfer in the field of safety-critical systems. It is an independent, non-profit organisation which cooperates with all bodies connected with safety-critical systems.

History
The Club was inaugurated in 1991 under the sponsorship of the UK's Department of Trade and Industry (DTI) and the Engineering and Physical Sciences Research Council (EPSRC), and is organised by the Centre for Software Reliability (CSR) at the University of Newcastle upon Tyne. Its Co-ordinator is Felix Redmill of Redmill Consultancy.

Since 1994 the Club has been self-sufficient, but it retains the active support of the DTI and EPSRC, as well as that of the Health and Safety Executive, the Institution of Electrical Engineers, and the British Computer Society. All of these bodies are represented on the Club's Steering Group.

What does the Club do?
The Club achieves its goals of technology transfer and awareness raising by focusing on current and emerging practices in safety engineering, software engineering, and standards which relate to safety in processes and products. Its activities include:

- Running the annual Safety-critical Systems Symposium each February (the first was in 1993), with Proceedings published by Springer-Verlag;
- Putting on a number of 1- or 2-day seminars each year;
- Providing tutorials on relevant subjects;
- Publishing a newsletter, *Safety Systems*, three times each year, in January, May and September.

How does the Club help?
The Club brings together technical and managerial personnel within all sectors of the safety-critical community. It facilitates communication among researchers, the transfer of technology from researchers to users, feedback from users, and the communication of experience between users. It provides a meeting point for industry and academe, a forum for the presentation of the results of relevant

research projects and technology trials, and a means of learning and keeping up-to-date in the field.

The Club thus helps to achieve more effective research, a more rapid and effective transfer and use of technology, the identification of best practice, the definition of requirements for education and training, and the dissemination of information.

Membership

Members pay a reduced fee (well below a commercial level) for events and receive the newsletter and other mailed information. Without sponsorship, the Club depends on members' subscriptions, which can be paid at the first meeting attended.

To join, please contact Mrs Joan Atkinson at: Centre for Software Reliability, University of Newcastle upon Tyne, NE1 7RU; Telephone: 0191 221 2222; Fax: 0191 222 7995; Email: csr@newcastle.ac.uk

CONTENTS LIST

SAFETY REQUIREMENTS

An Integrated Approach to Dependability Requirements
Engineering
I. Sommerville 3

HUMAN FACTORS

Integrating Human Error Management Strategies Throughout the
System Lifecycle
G. Burrett and S. Foley 19

Measuring and Managing Culturally Inspired Risk
*M. Neil, R. Shaw, S. Johnson, B. Malcolm, I. Donald
and C. Qiu Xie* 43

Safe Systems: Construction, Destruction and Deconstruction
J.M. Armstrong and S.E. Paynter 63

THE SAFETY CASE - 1

Developing a Safety Case for Autonomous Vehicle Operation on
an Airport
J. Spriggs 79

Managing Complex Safety Cases
T.P. Kelly 99

DEVELOPMENT AND LEGAL ISSUES

White Box Software Development
D. Daniels, R. Myers and A. Hilton 119

Reforming the Law on Involuntary Manslaughter: The
Government's Proposals on Corporate Killing
P. Popat and R. Eastman 137

THE SAFETY CASE - 2

Electronic Safety Cases: Challenges and Opportunities
T. Cockram and B. Lockwood 151

Safety Case Categories - Which One When?
O. Nordland 163

SAFETY ASSESSMENT

An Assessment of Software Sneak Analysis
G. Jolliffe and N. Moffat 175

Processes for Successful Safety Management in Acquisition
R.F. Howlett 189

Assurance of Safety-related Applications on a COTS Platform
C.H. Pygott 201

APPLICATION AND DEVELOPMENT OF STANDARDS

The Application of BS IEC 61508 to Legacy Programmable
Electronic Systems
H.M. Strong and D.C. Atkinson 217

The Software Safety Standards and Code Verification
I. Gilchrist 251

Evolution of the UK Defence Safety Standards
J.A. McDermid 261

AUTHOR INDEX 275

SAFETY REQUIREMENTS

An Integrated Approach to Dependability Requirements Engineering

Ian Sommerville

Computing Dept., Lancaster University, LANCASTER LA1 4YR, UK.
E-mail: is@comp.lancs.ac.uk

Abstract. This paper discusses an approach to system requirements elicitation that integrates safety requirements elicitation and analysis with more general requirements analysis. We propose that the analysis should be organised round pervasive 'concerns' such as safety and security which can drive the requirements engineering process. The paper introduces the notion of concerns based on business goals and discusses how concerns are used to elicit information about system requirements from stakeholders. I also discuss briefly how concerns may be part of a more general requirements engineering method called DISCOS that integrates requirements engineering with high-level design. I use examples from a medical information system to illustrate how concerns may be used.

1 Introduction

The discipline of safety requirements engineering is well-established in industries such as the chemical industry and the nuclear industry where safety issues have been paramount for many years. Conventionally, for computer-based systems, safety requirements engineering is considered to be a separate process from more general system requirements engineering with safety requirements being derived either before or in parallel with more general system requirements. This is embodied in, for example, the IEC 61508 process [1] [2] which was designed to support the safety engineering of protection systems.

The requirements of a protection system are distinct from the requirements of the system being protected so there is some rationale for separating the safety analysis of these systems from more general requirements analysis. The approach may also be applied in a more general system requirements engineering process for critical systems. It focuses attention on the importance of safety and may derive critical requirements that take precedence over other functional and non-functional system requirements.

The identification of safety requirements usually follows a process of hazard and risk analysis. In this process, hazards are identified along with the risks of these hazards arising, the risks of an associated accident and the potential consequences of such an accident. Various techniques may be used to support hazard analysis such as HAZOPS [3] and fault-tree analysis [4] and these can be the basis for deriving safety requirements.

However, the separation of safety and system requirements engineering can lead to problems. There are three related difficulties with this separation:

1. It assumes that safety is a system property that can be considered in isolation from other system properties such as integrity and timeliness.
2. It assumes that safety requirements can be isolated and clearly identified.
3. It increases the difficulty of identifying requirements conflicts and the costs of resolving these conflicts.

Separating safety analysis from requirements analysis suggests that we clearly understand what is meant by 'safety' for the system being analysed. From a common sense point of view we can say that a safe system is one that does not cause damage to people or its environment. However, when we try to pin this down into a more precise definition that can be used as a basis for assessing a system we run into problems. A train which does not move or a turbine that does not turn is clearly safe but of little use. A critical information system that provides correct information but in a font that can only be read with difficulty by operators with normal eyesight may not, in itself, be unsafe but can lead to safety-related failures in the encompassing socio-technical system. Delays in delivering an information system for cancer screening as a result of problems in reconciling safety requirements and other requirements may mean that several patients die unnecessarily. Availability, usability and timeliness of delivery all may affect whether or not the system is 'safe' – safety is not a simple system property that can or should be isolated.

Because safety is a holistic system property, I am convined that the notion of a 'safety requirement' as a special type of requirement that is distinct from other system requirements is potentially dangerous. While it may be applicable in the limited context of protection systems, I argue that it is more effective to consider safety as a pervasive 'concern' for the procurers, developers and operators of critical systems. Safety must be considered along with other concerns and that these concerns should drive the requirements engineering process for critical systems. While there may be specific requirements that focus on safety issues in a system, all system requirements can potentially affect system safety.

In the remainder of this paper, I develop this notion of concerns that reflect safety and other dependability properties and outline an approach to requirements elicitation and analysis where concerns structure this process. By using generic and specific questions associated with concerns, requirements can be elicited from system stakeholders. To illustrate the approach, I use examples drawn from a case study of an information system used to help manage the care of patients with mental health problems. In the final section of the paper, I briefly discuss a requirements and design method that uses concerns and that integrates system requirements engineering and high-level system design.

2 The MHCPMS system

In this section, I briefly introduce an example system that I draw on as a source of examples throughout the paper. This system is intended to help manage the care of patients suffering from mental health problems who may attend different clinics within a region. This based on a real system that is in use in a number of hospitals. For reasons of commercial confidentiality, I have changed the name of the system and have not included information about any specific system features.

The overall goals of the MHCPMS system are twofold:

- To generate management information that allows health service managers to assess performance against local and government targets.
- To provide medical staff with timely information to facilitate the treatment of patients.

The health authority has a number of clinics that patients may attend in different hospitals and in local health centres. Patients need not always attend the same clinic and some clinics may support 'drop in' as well as pre-arranged appointments.

The nature of mental health problems is such that patients are often disorganised so may miss appointments, deliberately or accidentally lose prescriptions and medication, forget instructions and make unreasonable demands on medical staff. In a minority of cases, they may be a danger to themselves or to other people. They may regularly change address and may be homeless on a long-term or short-term basis. Where patients are dangerous, they may need to be 'sectioned' – confined to a secure hospital for treatment and observation. These factors mean that safety is one of the issues that must be considered in the development and operation of this system.

Users of the system include clinical staff (doctors, nurses, health visitors), receptionists who make appointments and medical records staff. Reports are generated for hospital management by medical records staff. Management have no direct access to the system.

The system is affected by two pieces of legislation (in the UK, Acts of Parliament). These are the Data Protection Act that governs the confidentiality of personal information and the Mental Health Act that governs the compulsory detention of patients deemed to be a danger to themselves or others.

3 Concerns

Systems exist in an organisation to help that organisation deliver its organisational goals. We know that many systems that are developed are unused or fail to meet expectations. One reason for this is that the requirements for these systems either conflict with or are irrelevant to the organisational goals and constraints. Consequently, a good requirements engineering process should relate requirements to organisational goals and constraints.

Organisational goals reflect reflect the overall purpose and priorities of the organisation so the goals of a hospital (for example) are derived from its purpose to treat people who are ill or injured. Examples of the goals of a hospital might therefore be:

- Provide a high standard of medical care for patients.
- Ensure that a high proportion of resources are spent on patient care.

As well as goals, organisations must operate within a set of externally imposed constraints. These constraints may be legal, governmental or social and reflect the environment in which the organisational operates. Examples of constraints on a hospital might therefore be:

- Operate within the funding budget as set by the local health authority.
- Provide monthly reports to government on numbers of patients treated.

Goals and constraints may be conflicting so trade-off decisions on how best to satisfy the goals while meeting the constraints are inevitable.

The need to satisfy organisational goals has resulted in the development of a number of goal-based approaches to requirements engineering [5-10]. These are based on refining vague objectives into concrete formal goals then decomposing these further into sub-goals until a set of primitive goals which can readily be expressed as system requirements has been derived. These approaches have the advantage that they expose different goals from different stakeholders and provide a structured approach to assessing alternatives.

However, the vague and abstract nature of organisational goals and constraints poses difficulties for goal-based approaches. The inherent messiness of the world means that a hierarchical decomposition of goals is inherently artificial. Furthermore, I believe that stakeholders prefer goals to remain loosely defined and hence resist detailed goal decomposition. Loosely defined goals allow for flexibility of interpretation in whether or not these goals have been reached. Consequently, I think that validating the goal-decomposition hierarchy is practically impossible Other problems with goal-based approaches are discussed in a good summary of these techniques in a web site maintained by Regev [11].

To address the problems with goal-based approaches, we have introduced an intermediate concept called a 'concern' which helps bridge the gap between organisational goals and the requirements for systems being developed to support that organisation. Originally, we proposed that concerns should be integrated with a viewpoint-oriented approach to requirements elicitation [12, 13]. However, we are now convinced that 'concerns' are more generally applicable and can be applied in conjunction with any systematic approach to requirements elicitation and analysis.

Concerns, as the name suggests, reflect issues that the organisation must pay attention to and which are central to its operation. Concerns are identified from organisational goals by asking 'What do we need to think about if we are to achieve goal X or meet constraint Y'. Notice that concerns are not about what to 'do' but rather are a way of explicitly identifying the key issues around a goal.

Concerns correspond to high-level strategic objectives for the system. They are established after discussion with strategic management and are first expressed at a very high level of abstraction. They are frequently common to applications within the same domain. In general, it should be possible to express concerns using a single word or phrase and to explain in one or two sentences how these concerns are linked to organisational goals.

Some of the concerns that affect systems being developed in a hospital and their link to organisational goals might include:

1. Safety – hospitals must ensure that patient care is safe; from a legal standpoint, hospitals must work within national health and safety legislation.
2. Information quality – information quality is important for patient care and for providing timely and accurate reports to government about the functioning of the hospital.
3. Staffing – recruiting and retaining high-quality staff is essential to deliver a high standard of patient care.

Of course, there are many more concerns in a hospital and, when considering computer-based system development, it is important to decide which concerns are relevant. For example, privacy is obviously central in a medical records system but much less significant in a system that schedules ward cleaning. Concerns should

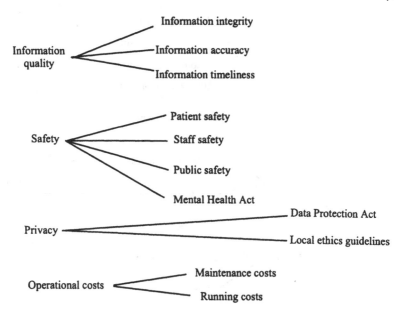

Figure 1 Decomposition of concerns in the MHCPMS

be ruthlessly prioritised so that only a small number of key concerns should drive the requirements engineering process.

For critical systems development, concerns are closely aligned with the dependability attributes that are most important for that system. Therefore, for a business-critical e-commerce system, the principal concerns are likely to be security and availability; for a control system for a radiation therapy machine, the principal concern is probably safety; for a telescope control system, the concerns may be reliability and accuracy.

Concerns, therefore, are a means of addressing the problems that arise when safety requirements elicitation is separated from more general systems requirements engineering. Where safety is important, it should be one of the concerns that drive the requirements engineering process. Safety issues can still be highlighted and analysed separately but the safety analysis is integrated with other analyses based around other concerns. By concern-cross checking, we can look for potential conflicts between requirements corresponding to the different concerns and can consider how other system requirements may have safety implications.

In the MHCPMS system, we have identified the principal concerns as:

1. Safety – the system should help to reduce the number of occasions where patients cause harm to themselves or others. The provisions of the Mental Health Act must be considered.
2. Privacy – patient privacy must be maintained according to the Data Protection Act and local ethical guidelines.
3. Operational costs – the operational costs of the system must be 'reasonable'.
4. Information quality – the information maintained by the system must be accurate and up-to-date.

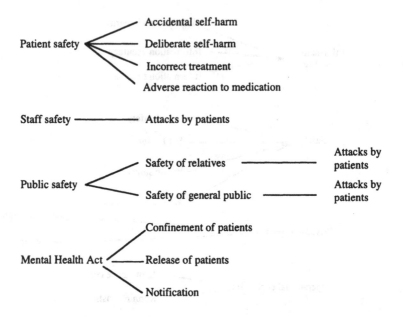

Figure 2: Further decomposition of the safety concern

High-level abstract concerns are decomposed into sub-concerns which reflect different facets of the concern. The first level decomposition of concerns into sub-concerns for the MHCPMS is shown in Figure 1. From these sub-concerns, a set of questions are derived. The outcome of the concern decomposition process is a set of questions, grouped by concern which are used during the requirements elicitation and analysis process. These questions are used to elicit information from system stakeholders and, from this information, system requirements are derived.

At this level of decomposition, concerns are still vague reflections of issues that the organisation considers to be important. To break these down into more detailed concerns, we ask 'what are the issues' questions such as 'what are the issues around patient safety that are of concern for the system'. This results in a further level of decomposition as shown in Figure 2.

Patient safety concerns the health and well-being of the patient themselves. Two of these are generic to all medical situations namely incorrect treatment and adverse reactions to treatment. The other two are more specific to mental health situations where the often confused nature of patients can result in accidental self harm and, sometimes, deliberate self-harm to gain attention.

The nature of patients suffering from mental health conditions means that they may attack other people. Although the threat is the same for medical staff, relatives and the general public, the risks and the situations where attacks might take place are different. Consequently, these are identified as separate concerns.

Finally, the Mental Health Act is concerned with both the safety of the public and the rights of patients. Legal formalities have to be followed when patients are confined and released, confinement can only be for a limited time without further examination and various people have to be notified when a patient is confined and released.

The process of decomposition continues by asking 'what are the issues' questions until these become difficult to answer. So the issues around the safety of relatives and the general public are 'Attacks by patients' but further decomposition into, say, types of attack is not needed for requirements derivation.

At this stage, the decomposition of concerns switches from identifying sub-concerns to identifying questions that may be used to elicit a deeper understanding of these issues and possible system requirements. In the case of an information system such as the MHCPMS, there are a number of generic questions that can be used as a starting point for these questions:

- What information from the system relates to the sub-concern being considered?
- Who requires this information and when do they require it?
- How is this information delivered to users of the information?
- What constraints does this concern impose on the system?
- What are the consequences of failing to deliver this information?

These generic questions may be decomposed into more detailed questions for system stakeholders or may be sufficient on their own to gather information about system requirements.

Possible answers to these questions about deliberate self-harm are:

1. Information about previous history of self-harm or threats of self-harm made during consultations

2. Medical staff during consultations. The patient's relatives or carers.

3. Can be delivered directly to medical staff using the system. Must be delivered to relatives and carers via a message from the clinic.

4. No obvious constraints are imposed by this.

5. Failing to deliver the information may mean that a preventable incident of self-harm takes place.

As well as the generic questions that may be used to gather information from system stakeholders, sub-concerns may be decomposed into more specific questions. For example, if we consider the information accuracy concern, then the generic questions are inappropriate and more specific questions associated with the concern may be derived. For example:

- How can potential inaccuracies in manual records be detected?
- If clinical staff selected inputs from a menu of options (to avoid inaccurate inputs) what problems might this cause?
- Is accuracy in some parts of the record more critical than others?

Notice that system functionality is not normally a concern. Goals are usually unaffected by details of the functionality provided by a system. However, where there is a clear link then 'System functionality' could be identified as a concern. We assume that, as part of the requirements engineering process, information on functionality is elicited by asking questions such as 'what user tasks are supported by the system?', 'what information is needed to carry out these tasks?', etc.

2.1 Concern cross-checking

A generic problem in complex systems is requirements conflicts where different system requirements are mutually contradictory. Conflicts are inevitable because different system stakeholders have incompatible goals and because of the interactions between the overall organisational goals and the constraints imposed on the organisation. Ideally, conflicts should be identified at an early stage of the requirements engineering process and resolved through negotiation. In practice, however, conflicts can be subtle and difficult to find in the detail of the system requirements. They may only emerge at later stages of the development process with the consequence that requirements changes and consequent rework become necessary.

The notion of concerns provides a mechanism that partially addresses the problem of detecting requirements conflicts as it allows cross-checking to be carried out at a higher level of abstraction than the requirements themselves. Rather than looking for conflicts in the requirements, we can use the concerns and our general background knowledge of these concerns to discover areas of potential conflict.

As there should be a relatively small number of concerns we can compare them in pairs to assess whether or not conflicts are likely to arise. For example:

1. *Safety and information quality.* Safety is dependent on accurate information. Conflicts are only likely if requirements on information quality allow records that are known to be erroneous or out-of-date to remain in the system.

2. *Safety and privacy.* Redundancy is an important mechanism for achieving safety but this requires information sharing. Privacy may impose limits on what information can be shared and who can access that information.

3. *Safety and operational costs.* There is always a trade-off between costs and safety so there is a potential for conflict here.

All we have done here is to highlight areas where conflicts are most likely and we need to turn to more detailed decomposition of the concerns to questions. We then can explore them in more detail.

Consider the privacy concern and its sub-concern of the Data Protection Act and the possible answers to the above generic questions about information, information users, information delivery, constraints and consequences of non-delivery. Possible answers to the generic questions identified in the section above are:

1. All information in the system that relates to identifiable individuals is covered by the data protection act.

2. All staff using the system need to be aware of the requirements imposed by the data protection act.

3. There are no information delivery requirements generated from this concern.

4. The constraint imposed is that personal information may only be disclosed to accredited information users where information users are people who need to do the information to do their jobs such as doctors or nurses.

5. Failure to address this concern could result in legal action being taken by data subjects.

The starting point for concern cross-checking is the constraints that are identified with each concern. We take these constraints and look at the relationship between them and the answers to questions generated in other concerns. Here we see an immediate conflict between the privacy constraints imposed by the DPA and the safety issue of making information about the possibility of self-harm known to the patient's relatives and carers (and possibly some medical staff). To improve safety, it makes sense to tell the patient's relatives about the possibility of self-harm. However, the patient may not wish their relatives to know of previous incidents so any dissemination of this information is not allowed by the DPA. There should not, therefore, be a requirement to generate information for patient's relatives included in the system.

2.2 Requirements derivation

Requirements are derived from the answers to the concern questions that are provided by system stakeholders. There is not a simple 1:1 relationship between the answers and requirements and it is up to the analyst to assess the answers and generate requirements from them. These should then be taken back to the stakeholders for validation.

Some examples of requirements that might be generated for the MHCPMS are:

1. The system shall provide fields in each patient record which allow details of incidents or threats of deliberate self-harm to be maintained.

2. The records of patients who have a history of deliberate self-harm shall be highlighted when accessed to bring them to the attention of clinical system users.

3. The system shall have a facility to generate e-mails to other accredited medical staff that warn about at-risk patients who may harm themselves deliberately.

4. The system shall only allow the transmission of personal patient information to accredited staff and to the patient themselves.

By using the answers to questions to generate requirements, we can avoid the problem of system stakeholders generating requirements that are too specific or which reflect their pre-conceptions about how the system should be designed.

4 Concerns and hazard analysis

The notion of concerns set out here has been illustrated using an system where hazard-driven analysis is not the norm. However, the concerns-based approach is completely consistent with hazard analysis for deriving safety requirements. To use concerns in conjunction with hazard analysis, the safety concern is decomposed as before into sub-concerns. However, after the principal sub-concerns have been

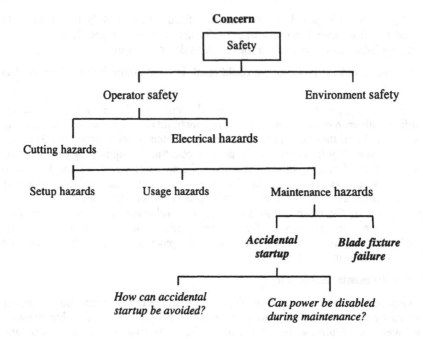

Figure 3: Concerns as hazards

identified, rather than decompose the concern into questions, each sub-concern is may be decomposed into a set of related hazards.

This is illustrated in Figure 3 which shows a hazard decomposition for a paper guillotine. You can see from this diagram that each identified hazard is represented as a sub-concern of safety and that the questions associated with concerns are designed to elicit information about how to avoid the hazard or mitigate the consequences of an accident.

5 Using concerns in a requirements engineering process

The notion of concerns as a driver for the requirements engineering process was first set out by us in a number of papers that described work on a requirements engineering method called Preview [12, 13]. Since then, we have developed a new requirements engineering method called DISCOS where we have extended the notion of concerns and have integrated early conceptual design with the process of requirements elicitation and analysis.

In the DISCOS method, we propose a spiral approach to requirements engineering as shown in Figure 4. Each stage in this process is:

1. *Define concerns*. As discussed here, establish the main concerns that influence the system and decompose these to sub-concerns and questions.

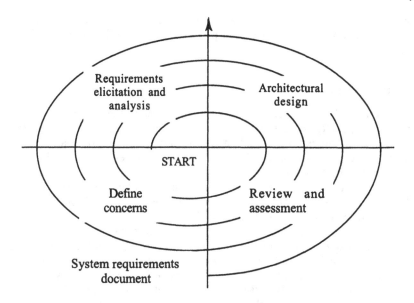

Figure 4: The spiral model of the DISCOS process

2. *Define system requirements.* Using the questions as a basis for the requirements elicitation, collect answers from system stakeholders and use these to define the requirements for the system.
3. *Propose system architecture.* Define an architectural model for the system showing sub-systems and their interactions.
4. *Assessment and review.* Review the requirements and model. If these are incomplete or incorrect, enter a new round of the spiral where more information may be collected and review issues addressed.

We propose a spiral process as we believe that creative human activities such as requirements engineering are never sequential processes. Concern definition, requirements elicitation and analysis and systems architectural design are always inter-leaved. In practice, irrespective of concerns, some requirements may be fixed at a very early stage in the system definition process. Equally, decisions on the system architecture may be made to allow legacy systems to be integrated, to reuse architectures with known characteristics and to help structure the process of requirements definition.

During each round of the spiral, more or less information may be added for each stage. The initial round is, essentially, a baseline round where high-level concerns are established as well as any requirements that are pre-defined for the system and, perhaps, an outline architectural design. Further early rounds focus on decomposing concerns with requirements 'popping out' as these concerns are established. Later rounds of the spiral are more focused on detailed requirements definition. We do not separate requirements engineering and high-level system design. This reduces the probability of proposing requirements which cause conflicts at the design stage.

The notion of concerns is pervasive and critical to the DISCOS method. We have discussed here how concerns can support the elicitation of requirements but they are also used to structure the review and assessment activity. Using concerns, this activity can assess whether design proposals may conflict with organisational goals.

6 Conclusions

The essential message of this paper is that separating the processes of eliciting and analysing safety requirements from more general processes of requirements engineering is undesirable. Such a separation is likely to result in safety requirements that conflict with other requirements with rework required to resolve these conflicts. Furthermore, it tends to isolate safety rather than emphasise that it is an issue of universal concern. The notion of cross-cutting concerns derived from high-level organisational goals avoids this separation while still maintaining the possibility of separate risk-based safety analysis.

Our original work on concerns focused on control systems but, more recently, we have been exploring how to use this approach for critical information systems. Methods of risk and failure based safety analysis are rarely applicable to this type of system. System failure does not have immediate implications for safety but, as we have seen with the MHCPMS, the system design should take safety into account. By highlighting critical issues such as safety and security as concerns, we can ensure that they pervade the requirements elicitation process and are not considered after the system functionality has been established. We are therefore now focusing on the use of concerns to support the requirements engineering of critical information systems and are planning further experiments in this area to develop the DISCOS method.

7 Acknowledgements

This research has been partially supported by the European Community's Framework Programme of IT research in the BANKSEC project (IST-1999-20711) where the DISCOS method was developed as project deliverable WP2.D3. For more information see http://www.atc.gr/banksec.

References

1.   Redmill, F., *IEC 61508: Principles and use in the management of safety*. IEE Computing and Control Engineering J., 1998. 9(10): p. 205-13.

2.   IEC, *Standard IEC 61508: Functional safety of electrical/electronic/programmable electronic safety-related systems*. 1998, International Electrotechnical Commission, Geneva.

3.   Chudleigh, M.F. and J.R. Catmur. *Safety assessment of computer systems using HAZOP and audit techniques*. in *Proc. SAFECOMP'92*. 1992: Pergamon Press.

4.   IEC, *International standard 1025: Fault tree analysis*. 1990, International Electrotechnical Commission: Geneva.

5.   van Lamsweerde, A., R. Darimont, and P. Massonet. *Goal-Directed Elaboration of Requirements for a Meeting Scheduler*. in *Proc. RE'95*. 1995. York, England: IEEE Computer Society Press. 194-203.

6.   Anton, A.I. *Goal Based Requirements Analysis.* in *Proc. 2nd Int. Conf. on Requirements Engineering (ICRE'96).* 1996. Colorado Springs: IEEE Computer Society Press. 136-44.

7.   Dardenne, A., A. van Lamsweerde, and S. Fickas, *Goal-Directed Requirements Acquisition.* Science of Computer Programming, 1993. **20**: p. 3-50.

8.   Fickas, S., Van Lamsweerde, A., and Dardenne, A. *Goal-directed concept acquisition in requirements elicitation.* in *6th Int. Workshop on Software Specification and Design.* 1991. Como, Italy: IEEE CS Press. 14-21.

9.   Mylopoulos, J., L. Chung, and E. Yu, *From Object-oriented to Goal-oriented.* Comm. ACM, 1999. **42**(1): p. 31-7.

10.   Rolland, C., C. Souveyet, and C. Ben Achour, *Guiding Goal Modeling using Scenarios.* IEEE Trans. on Software Eng., 1998. **24**: p. 1055-71.

11.   Regev, G., *Goal-Driven Requirements Engineering Overview.* http://lamswww.epfl.ch/Reference/Goal/Default.htm. 2002.

12.   Sommerville, I. and P. Sawyer, *Viewpoints: principles, problems and a practical approach to requirements engineering.* Annals of Software Engineering, 1997. **3**: p. 101-30.

13.   Sommerville, I., P. Sawyer, and S. Viller. *Viewpoints for requirements elicitation: a practical approach.* in *Proc. Int. Conf. on Requirements Engineering.* 1998. Colorado. 74-81.

HUMAN FACTORS

Integrating Human Error Management Strategies Throughout the System Lifecycle

Gretchen Burrett and Susie Foley
National Air Traffic Services (NATS)
Bournemouth, UK

Abstract

The disproportionate contribution made by human error to overall system risk has been well documented over recent years. Despite this fact, safety risk management processes continue to be biased towards focusing on the risk posed by equipment failures over that of human error. In order to redress the imbalance, this paper outlines a number of risk assessment methods that explicitly consider and integrate the human into the safety risk management process. A framework advocating a common language and structure is also proposed to facilitate the integration of traditionally independent risk management processes. Using a standard framework improves integration of human error into safety risk management and enables the sharing, comparison and validation of human error data across all stages of the system lifecycle.

1. Introduction - Human Factors Risk Management

Human error is an important consideration in complex safety critical systems because it makes the most significant contribution to overall system risk (Reason, 1990). However, historically, safety risk management has focused on addressing the risk posed by equipment rather than that posed by the human element of the system. Evidence of the overwhelmingly disproportionate contribution human error makes to total system risk indicates that current safety risk management has struck the wrong balance (e.g. Kinney, Spahn, & Amato (1977) and Federal Aviation Administration (1990)). In order to be truly effective, safety risk management needs to redress the imbalance and explicitly consider human error at each stage of the safety risk management process.

Addressing the risks posed by human error presents challenges to traditional approaches to safety risk management. Many of these challenges arise because of the very nature of human error, which has numerous causes and can therefore be difficult to fully understand, predict, model or prevent. For example, these errors can be the result of:

- the type of task being performed;
- the conditions under which the task is performed;
- human frailties such as fatigue;
- team factors such as supervision; and
- even routine or malicious violation of procedures.

Nevertheless, a systematic and informed consideration of the human as part of the safety risk management process can provide significant risk reduction, even if all risks posed are unlikely to be fully alleviated. Explicitly identifying and managing human error risks throughout the development and operation of complex systems can have the knock on benefit of improving operational effectiveness.

As with equipment reliability, human reliability targets and predictions made during the development of complex systems can be informed by data collected about actual performance in operation. In addition, it is important to ensure that claims made as part of safety cases remain valid once a system becomes operational. To date, some success has been achieved in "closing the loop" between predicted equipment safety levels and safety achieved operations. This process is equally, or possibly more important for the human element of the system. However, this is rarely achieved. The same situation existed for the safety risk management of hardware in the early days of reliability modelling, but this has improved over time.

Therefore, this paper will address the following two topics:

- Firstly, closing the loop between all stages in the safety risk management process requires a framework in which data used by relevant activities throughout the lifecycle can be shared and compared. This paper will propose a framework to link key safety risk management processes in a way that facilitates management of equipment and human safety risks; and
- Secondly, methods for explicitly considering and integrating the human into the safety risk management process will be outlined.

2. A Framework to Link Error Management Processes

This section addresses the goal of closing the loop between all stages in the safety risk management process. To achieve this, a framework must be established to ensure a common language and structure. It should address the following objectives:

- Consider human and machine contributions to system success and risk explicitly, and alongside one another, to improve understanding of the balance of risk and prioritisation of risk reduction activities;
- Represent the complex nature of human error and its many causes;
- Set and measure adherence to safety targets at each step of the lifecycle;

- Improve our ability to set targets through better operational performance data, as well as validating predictions against operational reality;
- Include new or modified sub-systems (which may be added at different times) into safety risk management;
- Facilitate trade-offs between risk reduction options; and
- Assess the effectiveness of risk reduction measures.

The framework will need to link many key safety risk management processes from throughout the system lifecycle. Those considered in this paper are as follows:
- Identification of risk reduction strategies;
- Setting of risk reduction targets;
- Predictive error management processes;
- System design process;
- Development of Safety Cases;
- System testing and evaluation; and
- Monitoring actual safety performance in operations.

A variety of methods currently exist to achieve the above processes (e.g., Risk registers, Failure Modes, Effects and Criticality Analysis (FMECA), HAZOPs, structured incident investigation techniques). To enable risk reduction processes to be effectively interlinked, this paper advocates the development of a common structure and language, such that information and data can be shared and compared at each stage of the lifecycle.

The primary framework elements proposed are listed below:
- **Goals** – The safety, operational and business aims of the system.
- **Functions** – The actions that need to be performed (by either the human or machine component of the system) in order to achieve the system goal.
- **Failure Modes** – The human or equipment failures that could result in the goal *not* being achieved.
- **Causal Factors** – A set of factors that describe how the functions can result in error (e.g. lack of training, poor design, fatigue).
- **Conditions** – the range of conditions under which the goals must be achieved (e.g. 24 hours a day, in poor weather, in a defined working environment)
- **Mitigations** – System defences in place to prevent the incidence/ reduce the consequences of failure modes (e.g. procedures, error checking, supervision)

These factors are generic in the sense that they could be defined in a language that remains constant throughout all stages of the lifecycle. From this it is possible to use related concepts and data throughout the system lifecycle. For example, targets set to reduce a particular causal factor by a specific amount can be made during the requirements phase and throughout the design of the new system. If the same term is used to collect safety data during operation of the system, then comparisons can be made to determine the extent to which the target has been achieved. The assumptions made about the conditions under which the system functions would be carried out can be compared with the conditions under which system errors are occurring.

The framework would be continually updated as a result of the information generated by its use throughout the various risk management processes. Thus the framework could provide a mechanism for sharing the output and lesson learning between all of the previously disparate processes as well as a means for increasing efficiency by preventing duplication of effort. Figure 1 below, derived in part from Reason (1990), illustrates the proposed framework.

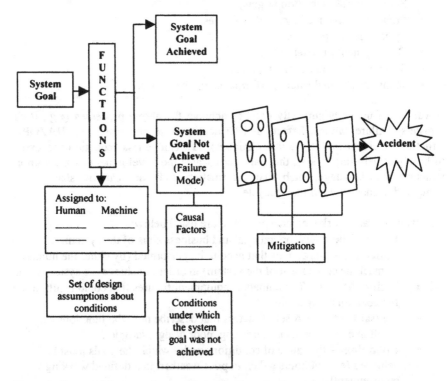

Figure 1. Proposed framework to enable integration of safety risk management processes.

Any safety critical system has an overall system goal. For example, in air traffic control, the fundamental safety goal of the system is maintaining safe separation of aircraft. In a nuclear processing system, the overall goal may be to produce nuclear power safely and efficiently.

A set of specific functions must be performed within a target level of safety in order to achieve the system goal. Performance targets can be assigned to both functions and goals to define safety targets/requirements. During system design, the functions may be assigned to either the human or the machine to perform. System designers make a set of assumptions about the conditions under which the functions will be performed. An example of an ATC function would be the detection and resolution of potential conflicts between pairs of aircraft.

If both the human and machine perform the functions in accordance with the design intent, the system goal is achieved. If either the human or machine does not perform the functions in accordance with the design intent, the system goal is not achieved and an error or failure occurs. An example from ATC would be issuing the wrong clearance to an aircraft. Both the type and frequency of failure modes are important to safety risk management. A measure of the criticality of a given failure can be determined by the extent to which it relates to *not* achieving key safety goals. The framework advocates the use of a common set of causal factors and conditions under which the system failure took place.

The adverse event is prevented from turning into an accident via a series of mitigations that are in place to act to block its path. Mitigations are divided into three main types:
- those that reduce the likelihood of the error taking place;
- those that aid error detection; and
- those that aid error recovery.

Mitigations can involve the human (e.g. training and supervision), the machine (e.g. alert devices) or procedures (e.g. procedures designed to mitigate risk). However, as Reason explains (1990), mitigations (system defences) have limitations or 'holes'. They are not perfect in blocking the error and when the holes in the mitigations line up, error is allowed to continue along its path to an accident. Tracking the predicted and actual performance of mitigations can be central to ensuring that safety risks are managed effectively.

Traditionally, safety risk management effort has focused on preventing those failures that could result in accidents. However, this framework can also be used to look on the positive side to examine those functions that increase the likelihood or the extent to which safety goals can be achieved.

3. The Process

The following sub-sections outline the process to be followed in putting the framework into practice to manage safety risks. The process described in this section is an iterative loop. For the purposes of this paper, we will begin our discussion of the process during operation, where the first steps are to understand the current system risks and then to identify strategies for reducing those risks. Next, risk reduction targets need to be set so that there are clear goals for where system risk should be in the future. There needs to be a proactive attempt to identify the opportunities for error (and their mitigations) that may arise during the development of system improvements. Throughout the design of any improvement to the system, the development of its safety case and its testing and evaluation, the anticipated performance needs to be compared against the risk reduction targets. Once in operation, the actual performance of the system needs to be measured against the targets set. This needs to be a continuous process of stepped risk reduction. Figure 2 below illustrates the process. The following sections focus on mechanisms to apply the integrated framework, and also to explicitly consider the human, at each stage in the process.

3.1   Understand Current System Risk

The continuous improvement of operational systems requires an effective performance monitoring system, which can identify problems, their relative importance and potential solutions.   This understanding of current system performance and risk provides a baseline from which system improvements (e.g., training, procedures, new equipment) to reduce risk can be identified, developed and implemented. Key elements in building this understanding include:

- An effective incident investigation process;
- A proactive safety risk reporting process;
- An open reporting culture;
- A mechanism for tracking trends (in failure modes, causal factors, conditions and mitigations);
- A process for using trend data in combination with risk/benefit/cost data to prioritise action.

Incident Investigation Process

At the forefront of the identification of system risk is a robust incident investigation process that identifies risks posed by equipment, procedural and human failures evident from incidents or accidents. The investigation process should include an analysis of specific issues by experts in that area. For example, an error analysis by human factors experts where human error is involved, an analysis by technical equipment experts where equipment failures occur and an analysis by operational experts, such as air traffic controllers in the case of the ATC system, where procedures are involved. An important benefit of applying the framework to incident investigation and analysis is that these specialists all use a common structure and language, which can be more efficient and productive.

Once methods are in place for capturing risks, it is vital to ensure there are mechanisms to truly understand the nature of the risks posed. A robust incident investigation process needs to determine how errors and incidents are occurring, what is causing them, those that are critical, the conditions under which they are more likely to occur and the effectiveness of mitigations for human and equipment failure modes.

The incident investigation process needs to be structured in such a way as to go beyond describing what happened and determine why it happened. Establishing the reasons why an operator made an error is vital to understanding which interventions might prevent or reduce the likelihood of reoccurrence.

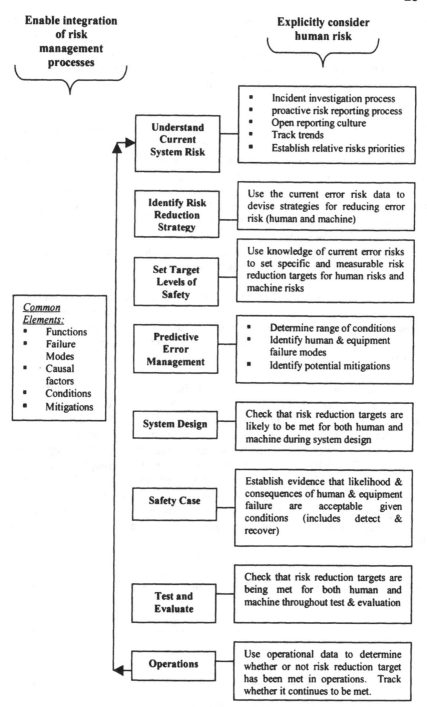

Figure 2. The risk management process

'Human error' is no longer an acceptable level of detail to go to in determining the causal factors of an incident. Vague descriptions of human error do not enable effective solutions to human error risks to be identified. The human needs to be included explicitly in the incident investigation process if there is to be any hope of tackling the largest contributor to overall system risk in safety critical systems today.

Additionally, data on the conditions under which errors and incidents occur should be systematically collected. Humans do not act in a vacuum; rather they are influenced by the conditions under which they operate. From many accident and incident reports in a variety of industries, it is apparent that accidents result when active failures (like human error) combine with existing latent conditions lying dormant within the system. It is essential that the incident investigation process rigorously collects and classifies these conditions that increase safety risk in addition to the key causal factors implicated in incidents. Through developing an understanding of these error-provoking conditions, an additional opportunity to reduce the risk posed by human error is presented.

In order to obtain a level of error detail that actually enables error problems to be clearly identified and understood, a robust error classification technique is essential. Whichever specific technique is used, it is vital that it enables the consistent collection of key error detail using a standardised set of classifications in order to enable an accurate picture of error risk to be obtained.

The Technique for the Retrospective (and predictive) Analysis of Cognitive Error in air traffic management (TRACEr) is one such technique developed by NATS (Shorrock, 1997). TRACEr is based on models of human information processing where human error is viewed as a failure of human information processing. Figure 4 shows a very basic human information processing model (adapted from Wickens, 1992) in which the human perceives information through any of the five senses and then judges, plans or decides what to do with that perceived information. In deciding what to do, the human accesses both long-term and working memories for relevant information to inform the decision-making process. Once a decision is made, the human then takes action to carry out the decision.

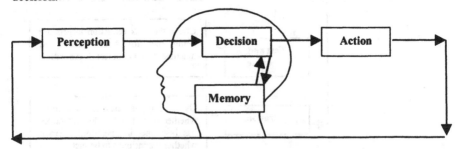

Figure 3. Human information processing model (adapted from Wickens, 1992).

Human error can arise from a breakdown in any of these processes. For example:

- **Perception** - misperceive or fail to perceive information correctly.
- **Decision** -an error of judgement, planning or decision–making can occur.
- **Memory** - information can be forgotten or misrecalled.
- **Action** - all of the information processing can be successfully carried out (correctly perceived, memory accessed correctly and the right decision made), but an error can occur in carrying out the action.

By considering human error in these explicit terms and collecting human error detail that describes which aspect of human information processing failed, how it failed and why; the process of identifying human error interventions that may prevent or mitigate specific errors becomes more informed. It is important to know how and why the information processing failed because it is only with this level of detail that effective interventions to specific error risks can be identified. For example, again from ATC, a controller may issue an incorrect clearance to an aircraft. This could be caused by an error in perception (e.g. misreading the flight progress strip), a decision error (e.g. expecting the aircraft to achieve a rate of climb outside its normal performance), a memory error (e.g. forgetting that another aircraft was already at that level) or an action error (e.g. simply saying climb to flight level 110 instead of flight level 100). Clearly, the solution to the error would be significantly different depending on the actual cause.

TRACEr uses a standardised series of pick lists and decision trees to enable consistent classification of the following error information for each area of information processing (see Tables 1-4).

Error Mode	Error Mechanism
Perception	
Mishear Mishearing a readback or transmission.	**Expectation** Perceptual errors driven by expectations.
Mis-see Misread, misperceive, or misidentify visual information.	**Confusion** Misidentify/ misperceive due similar/ confusable of appearance/ spatial position.
No detection (auditory) Failing to detect, or being late to recognise the significance of, a readback or transmission.	**Discrimination failure** Failing to see or hear something that is vague or of short duration.
No detection (visual) Failing to detect or identify visual information, or detecting information too late to be effective.	**Tunnel vision** Fixating, tunnelling or 'black-holing' to the exclusion of other relevant information.
	Overload A large amount of incoming information.
	Distraction / Preoccupation Momentary distraction/ long-term preoccupation.

Table 1. Did the controller/pilot mis-see or mishear, or fail to see or hear something?

Error Mode	Error Mechanism
Memory	
Omitted or late action	**Confusion**
Forget to perform a planned task, or miss a step in a task sequence (e.g. monitoring information/people).	Other (e.g. similar) information interferes with memory.
Forget information	**Overload**
Forget information or previous actions.	Too much information to retain in memory.
Misrecall information	**Insufficient learning**
Misrecall temporary or longer-term information/actions.	A learning problem or negative transfer of information.
	Mental Block
	A mental block - just cannot recall information.
	Distraction / Preoccupation
	Momentary distraction or longer-term preoccupation.

Table 2. Did the controller/pilot forget or misrecall information, or forget to do something?

Error Mode	Error Mechanism
Decision Making	
Misprojection	**Misinterpretation**
Misprojecting or misjudging spatial-temporal information in trying to maintain separation.	Failure to integrate, calculate or understand information.
Poor decision or poor plan	**Failure to consider side- or long-term effects**
Poor decision or inadequate plan.	Unforeseen side- or long-term effects.
Late decision or late plan	**Mind set**
Acceptable decision or plan formed too late to be fully effective.	Sticking to a faulty plan, belief or interpretation, despite evidence to the contrary.
No decision or no plan	**Knowledge problem**
No decision made or no plan formed for an aircraft.	Lacks required knowledge due to training or learning.
	Decision freeze
	Decision 'freeze' due to complexity or emotion.

Table 3. Did the controller/pilot make an error in projecting, planning or decision making?

Error Mode	Error Mechanism
Action	
Selection error Unintended manual selection or positioning.	**Variability** Lack of manual precision, fluency or intonation.
Unclear information Transmitting or recording unclear, vague or ambiguous information.	**Confusion** Selecting an object that looks similar to another, is in a confusable position, or is functionally similar.
Incorrect information Inadvertently transmitting or recording incorrect information.	**Intrusion** Thoughts or habits cause a controller to do or say something unintended.
	Distraction / Preoccupation Momentary distraction or longer-term preoccupation.
	Other slip Other slip of the tongue, pen, action etc.

Table 4. Did the controller/pilot perform an action in a way not intended, or inadvertently say something that was incorrect or unclear?

TRACEr also includes a standard set of Performance Shaping Factors (PSFs) which can be used to classify the conditions under which the incident occurred. PSFs do not cause incidents, but rather are factors that increase the likelihood that an incident may occur. TRACEr groups PSFs under nine key headings to enable the key areas of concern to be highlighted through data collection:

Performance Shaping Factor (PSF)	
Social and Team factors	
Allocation of function and responsibility	Team pressure
Team relations	Sector manning
Personal factors	
Alertness/fatigue	Confidence
Emotional or occupational stress	Job Satisfaction
Procedures	
Number	Comprehensiveness or completeness
Complexity	Duration in use or stability
Training and experience	
Task familiarity	Mentoring
Level of experience	Time on sector
Workplace design, HMI and equipment factors	
Console (or flight deck) ergonomics	Radar
Electronic tools	Equipment

Performance Shaping Factor (PSF)	
Ambient environment	
Noise	Temperature
Lighting	Air quality
Pilot-controller communications	
RT workload	Controller RT standards
Pilot language or accent	Pilot RT standards
Traffic and airspace	
Traffic load	Traffic complexity
Weather	Sector design

Table 5. Performance Shaping Factor (PSF) examples

The collection of incident error information is critical. In order to reach potential solutions to human error problems, it is imperative to obtain high quality and specific error detail. It is vital that standardised and specific data collection techniques are used and that the quality of information submitted by reporters is assessed. Although resource intensive, post incident interviews represent an extremely effective way of ensuring that the critical error details from an incident are collected.

Proactive Safety Risk Reporting Process

In addition to using the reactive incident investigation and analysis route to identify risks, more proactive measures can be utilised. For example, safety observation reporting schemes and the reporting of potential incidents can be used to proactively identify risks before they result in adverse events. It is useful to develop a living "risk register" to collate and document an overall picture of system risk, on an ongoing basis, from all of these various sources. The same common failure mode (e.g., human error), causal factor and condition (e.g. PSF) structures must be used to enable data collected proactively to be linked with that collected by reactive incident analysis.

Open Reporting Culture

An open reporting culture, where the reporting of potential incidents and safety observations is actively encouraged, is a prerequisite to the proactive identification of current risks. The development of an open reporting culture is not the subject of this paper. However, consideration needs to be given to managing possible concerns over blame or disciplinary action for individuals who report their errors. Consideration also should be given to providing feedback to individuals and groups on actions taken as a result of "voluntary" safety reporting.

A Mechanism for Tracking Trends

Whether data is collected reactively or proactively, it is essential to collect this data in a consistent manner to ensure its integrity and enable reliable trend data to be elicited. In this way, the risk will have been identified, the reasons behind it's occurrence will have been identified as best as possible and it will be known whether or not the risk is posing an increasing threat to the system. It is only through a standardised format for expressing human error detail that important trends will be allowed to emerge. Figure 4 depicts an example of trend data arising from Performance Shaping Factors or conditions. The figure indicates that, although it is the most frequently arising factor, the trend is for training factors to be on the decline. This perhaps indicates that interventions are having a successful impact. In contrast, environmental factors appear to be worsening over time and design factors appeared to being staying constant.

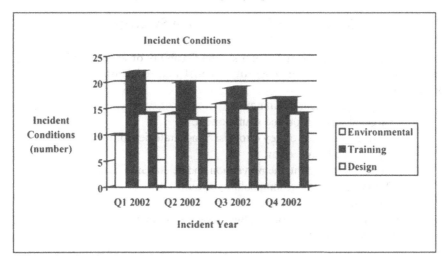

Figure 4. Trend data on Performance Shaping Factors

By collecting error information in a structured way, it is possible to elicit trends and to link the incident investigation process with other error management processes throughout the system lifecycle. In organisations managing large safety critical systems, there may be many independent processes and individuals involved in investigating and analysing incidents. In order to obtain maximum usability of this vital 'lynch pin' data, it is essential that there is a commonality of language and definitions for causal factor information. For example, NATS has developed a common list of comprehensively defined causal factors in an attempt to achieve a common language to which anyone working to reduce risk can refer and clearly understand (e.g. system designers, operations staff, trainers, managers, procurement staff, personnel, safety case and safety management system staff).

In addition to systematically collecting information on the causes of incidents, it is necessary to understand which errors or failures result in the riskiest situations.

To achieve this, two factors will need to be considered: firstly, the frequency of each error type and secondly, the level of risk posed by resultant incidents. For example, in addition to an assessment of the causal factors and conditions associated with an incident, a separate assessment of the risk posed by the incident could be made. Such an assessment may consider a range of factors associated with the incident including the consequences of the incident, the stage at which detection and recovery took place and the safety barriers invoked, in reaching a risk classification.

Once the relative priorities for risk reduction have been determined, the current system risk information can be interrogated to try to identify system improvements to reduce specific risks. Ensuring high quality and a sufficient level of detail in the collection of current risk data facilitates the identification and development of relevant risk reduction measures.

3.2   Identification of a Risk Reduction Strategy

In identifying strategies for reducing risk, the feasibility of various options needs to be considered. A varied range of potential risk reduction strategies exists including:
- Targeted individual training;
- Targeted supervision and management;
- Raising awareness of key error risks and strategies for reducing risk;
- Training on human error;
- Future tools to aid in the prevention and detection of human error;
- Design changes (e.g. font, colour, layout);
- Changing procedures; and
- Sharing risk lesson learning throughout the organisation and with other similar organisations.

In considering the feasibility of various risk reduction options, a number of factors should be weighed against one another. For example, the likely risk reduction impact of the initiative should be determined. This should be defined in terms of target performance levels for key goals or functions. Data collected during operations can also be used to identify specific causal factors, conditions, or mitigations that are most likely to improve safety performance. The potential for new risks to be introduced as part of any change should be considered and associated mitigations should be identified. The likely costs of the initiative (including personnel related costs such as those incurred through retraining and additional supervision) should be established. Through a cost-benefit analysis, which considers the potential for new risks, it should be possible to determine the most feasible risk reduction options for specific human or equipment failure modes.

3.3 Set System Requirements (Target Levels of Safety)

Data on current system performance and risks, together with targets developed as part of the strategic planning, facilitate the development of more realistic system safety requirements earlier in the development of equipment, procedures and/or training. Defining measurable safety requirements at an early stage in development has three main advantages:

- Requirements can be set for total system (human + equipment + procedures + training) performance. This integrates consideration of the human into specifications, contracts and acceptance;
- This provides a framework in which predictions of likely human and equipment performance against the requirements can be used when assessing design options or making trade-offs; and
- It also helps to ensure that evidence of human and equipment contributions to achieving the requirements are formally recognised by systematic analysis and evaluation to build a safety case which give evidence that required performance levels will be met by the total system proposed.

As discussed in the introduction, one of the difficulties of including the human in safety requirements is that their performance can be dramatically affected by:

- The task they are performing;
- The level of performance required;
- The conditions under which they are operating (e.g., noise, climate, uncertainty);
- Personal factors (e.g. fatigue);
- Other people in the system; and
- The quality of equipment, procedures and training.

Therefore, requirements need to take these factors into account. This means that safety requirements are closely linked to operational performance requirements. As such, the key steps to achieving a defined and testable set of total system requirements are as follows:

- a) Identify top level statements of need which identify key goals, system performance and safety targets ;
- b) Understand the functions that need to be supported and the performance/safety requirements of each individual function;
- c) Prioritise functions so that their impact on safety and performance is clear and explicit;
- d) Develop scenarios which contain a representative set of functions, conditions and events against which the resulting requirements can be tested; and
- e) Be clear about the evidence required to show that the requirement has been met.

In the first instance, steps a) and b) above should be independent of equipment or procedural solutions, because this enables more effective trade-offs about how well various solutions might enable the goal/function to be achieved. This is also fundamental for linking between lifecycle stages, as use of generic functions enables comparison between existing and future system performance levels. For future systems, human and equipment failure modes and mitigations will emerge as the design develops.

Steps c & d above are instrumental in limiting the scope of step e. It is important to have an understanding of the range and relative importance of functions that need to be performed by the total system in order to achieve the goal. In addition, there is a need to identify those operational conditions and events that conspire to compromise the ability of the human operator to maintain the required levels of performance. For example, weather conditions, time of day, unplanned events and equipment failures are all conditions that play a part in the performance of the human. Establishing these conditions is an important pre-requisite of establishing the scenarios under which the system must be capable of demonstrating its ability to meet performance and safety targets. It is also helpful to agree a set of baseline assumptions about representative levels or ranges of user experience, training and skill. Scenarios should also cover the need for effective team working to meet safety requirements.

Clearly, it is not practical or cost-effective to develop and/or test all possible combinations of conditions that could affect human performance. The aim should be to define a set of scenarios that characterise the range of functionality and conditions (particularly high priority functions and 'risky' combinations of conditions) in order to achieve sufficient risk reduction in subsequent analyses and tests.

Scenarios therefore need to be developed to ensure that the following elements are identified and tested:

- the full range of functions that must be performed successfully;
- the impact of functions that must be performed concurrently;
- the integration of functions that must be performed consecutively;
- the range of events that provide the greatest test of human intervention in operational events (i.e. safety related and/or non-standard tasks);
- the conditions under which human operators are expected to mitigate system failures; and
- team roles, co-ordination and cross-checking procedures and routines.

Whilst this is a difficult area, project experience suggests that waiting until the end of the project to find out how well the human-machine system meets the often loosely defined total system safety objectives is bad for both the customer and the supplier. Therefore, although the perfect solution may not exist, considerable risk reduction can be achieved by defining a set of total system performance requirements. This enables iterative assessment of how well proposed solutions meet the operational need, facilitates trade-offs and enhances the integration of sub-systems towards the overall system requirement.

The risk reduction targets which are set should become the yardstick against which all subsequent phases of the system lifecycle are continuously measured: from initial design through to the operational phase. Setting requirements in terms of goals and functional performance targets is a way of enabling traceability between data collected in operations, setting requirements and the data developed during design and testing of system improvements.

3.4   Predictive Error Management

In developing new or modifying existing systems (for example, to implement system improvements), the effects on system risk need to be predicted and assessed. Effective risk prediction begins at the earliest stages of system development (e.g. conception and design). At these early stages, risk predictions can be relatively imprecise. However, they are valuable because they provide an early warning of potential risks in the system enabling an improvement plan to be developed when the system is most malleable to change and when the cost of modifying the system is least prohibitive. Once in operation, even minor changes can have substantial cost implications.

As the system develops, its properties become more precisely known and the risks can be predicted with greater certainty. Continuous refinement of the accuracy of predictions throughout system development enables risks to be addressed at the earliest possible time. Early prediction enables modifications to be made to manage risk at a phase of the lifecycle when more options are available to design risk out of the system (or mitigate it more successfully).

One of the big differences between human error prediction and equipment failure prediction is the use of scenarios or use cases to represent the range of conditions (which, as we know, impact human performance). Therefore, scenarios developed as part of system requirements are an important input into error prediction. Additionally, the goal and functional performance criteria developed for the system requirement provide information on the accuracy and criticality of various activities, which is necessary to inform predictions. Data about those errors that cause safety problems in current operations is also a useful input to predictive error assessments, as it enables an assessment of how well proposed solutions reduce the likelihood of, or mitigate against, known errors. This identification of the potential safety benefits of a system improvement can be performed in conjunction with a predictive assessment of likely human error risk.

The key steps in performing error predictions include:
- Agree scenarios and functional performance criteria to be used in predictive error assessments;

- Determine likely allocation of function between the human and machine. This should result in one or more options for human tasks to be performed and grouped into roles for various members of the operational team. Predictive error assessments use this function allocation as a basis for systematically predicting where errors can occur.
- Use a structured method such as predictive TRACEr Lite (Shorrock, Kirwan, MacKendrick & Kennedy, 2001) and/or HAZOP (CIA, 1987) to systematically walk through scenarios and determine possible errors, their consequences and potential mitigations. Existing data on system failures (e.g. records on system reliability) and human error (e.g. incident data and safety observations) can be used to inform the predictions.

Table 6 shows an example of the output of a combined application of HAZOP and Predictive TRACEr to a single intention statement. The intention statement indicates the task that the human is intending to carry out. HAZOP consists of a series of guidewords that are combined with intention statements to highlight deviation opportunities. The possible causes of the deviation are recorded, as is the consequence. The HAZOP technique requires that the system indication, system defence and human recovery are explicitly detailed for each intention statement, thus enabling the identification of the hazard.

The Predictive TRACEr elements tackle the same intention statement from a different perspective, identifying opportunities for failures in each of the information processing domains (perception, memory, decision and action). From the application of both techniques to a series of intention statements, it is possible to produce a list of potential errors. The success of HAZOP in enabling the predictive identification of error risks has been well established since that method has been in existence since the 1960s. Although Predictive TRACEr has been developed only recently, preliminary indications of its capabilities are promising. One study found that Predictive TRACEr enabled the identification of 94% of human errors that later occurred during the simulation of the system (Shorrock, 1999).

HAZOP component	
Intention Statement	To **find** a predicted breach interaction using the tool.
Guideword	Late
Deviation	Controller finds the breached interaction too late
Cause (s)	- Distraction - Filtering will not show traffic which is not the present sector's responsibility, which may become own responsibility too late - Traffic not co-ordinated into sector
Consequence	Late or non resolution of conflict
System Indication	Interaction alert when first a/c transferred
System Defence	Interaction Alert at 5 minutes and 2 minutes.
Human Recovery	Stick to co-ordination procedures
HAZARD or Question	Keep to current procedures for clean transfer of a/c. Include instruction that a/c is not to be transferred if it

	appears in the tool.
TRACEr component	
Perception Error	- Information not perceived due distraction
	- Information not perceived due to filtering algorithm
	- Information not perceived due inaction by another controller
Memory Error	- Unintentional deviation from procedures
Decision Error	None
Action Error	None

Table 6. A combined HAZOP/ Predictive TRACEr output.

3.5 System Design

The design of equipment, procedures and training has a major influence on the likelihood and result of human errors. Therefore, failure to consider the impact of the design on the expected human performance can lead to a gross misjudgement of total system safety and operational effectiveness. Throughout the design phase of any new system or system improvement, the design should be regularly checked against safety requirements (which should include risk reduction targets) to ensure the system design is on track to enable those targets to be achieved.

As the preferred design emerges, it will be possible to allocate function between the human and machine and therefore identify human and equipment failure modes. At this stage, mitigation for human and equipment failures can also be developed and assessed. Data from current operations, together with predictions of human and equipment performance (see Section 3.4) can be used to assess the extent to which the proposed solution is likely to achieve system targets.

This type of assessment can be facilitated by a number of tools. For example,

- **A task analysis** - which describes tasks, information requirements, feedback, likely errors, detection and recovery procedures;
- **Scenarios and performance targets** – developed as part of the System Requirements for use in walkthroughs and proposed system assessments;
- **Human factors design criteria** – standards and guidelines for good design, which generally reduce the risk of human error;
- **Prototypes, mock-ups, and models of the system** – which enable users to 'test drive' the system and spot problems which increase errors or reduce the effectiveness of detection and recovery;
- **Developing procedures alongside the design** - to check that all sub-systems work together to support task performance. These procedures should also be assessed to identify risks associated with inaccurate or inappropriate application of the procedure; and
- **Defining training in parallel with the design** -this aids understanding of likely required skill and experience levels of personnel. It also allows assessment of the extent to which additional training to improve mitigation of failure modes is achievable within training budgets.

3.6   Safety Cases

The safety evidence required will depend upon project scope and the extent to which human factors risks are present. Identification of human factors risks should be included in the project risk register, which can be used at all stages of the lifecycle. This information should be gathered iteratively to ensure that issues are identified as early as practicable, to reduce the redesign time and expense.

The safety case should then include evidence that total system requirements have been met and that human error risks have been mitigated. The type of evidence required to address human error risks for most projects, together with the key techniques for providing that evidence, are summarised in the table below.

Required Evidence	Methods
Analysis that shows that all potential human errors have been identified.	Failure Mode Identification, Human Reliability Analysis (HRA), Fault Trees, HAZOP, Incident Investigation.
Analysis to identifies conditions/ events that increase the probability/ impact of human error.	Brainstorming, task analysis, human performance modelling & workload analysis, Incident Investigation.
Prioritisation of the safety significance of human errors, bearing in mind the potential for detection and recovery of those errors.	FMECA, HRA, HAZOP, Fault Trees, Safety Model, Incident Investigation.
When the human is to be used to mitigate known design deficiencies and/or system failure modes, evidence that safety levels will not be compromised.	FMECA, HAZOP, Fault Trees, Safety Model, HRA, task analysis, workload analysis, human performance modelling, human-in-the-loop prototyping/simulation, Incident Investigation.
Evidence that training needs have been identified and that expected training will support necessary levels of human performance.	TNA, task analysis, human performance modelling workload analysis, human-in-the-loop prototyping/ simulation, Incident Investigation.
Definition of operational procedures and evidence that human performance in defined operational scenarios using those procedures will be acceptable.	HRA, Fault Trees HAZOPS, Safety Model, task analysis, human performance modelling workload analysis, human-in-the-loop prototyping/ simulation, Incident Investigation.
Identification of operator & maintainer roles within teams, including supervision, monitoring & cross-checking of safety critical actions.	FMECA, Role definition, task analysis, workload analysis, human-in-the-loop prototyping/simulation, Incident Investigation.
Human-in-the-loop assessments of selected scenarios to verify results of the analyses above.	Human-in-the-loop prototyping/simulation, Incident Investigation.
Audit trail showing that human factors issues have been identified, integrated into relevant project activities & included in the design of equipment, procedures & training to mitigate safety risks.	Issues database, Hazard logs, project risk registers.

Table 7. Safety Case evidence and methods

3.7   Testing and Evaluation

During the testing and evaluation phases of system development, it is important to test system performance against the targets to determine whether or not they have been successfully achieved and therefore whether or not the system is on track to meet the safety and/or operational performance targets set out in the requirements. As far as practicable, the system's risk reduction performance should be tested and evaluated under all conditions for which the system was designed to determine whether or not the targets, and therefore the system requirements, have been met.

As with safety cases, the evidence required to establish that the system supports safe and effective operation by its intended users will depend upon project scope and the extent to which human factors risks are present. This evidence should be gathered iteratively, throughout the lifecycle of the system, to ensure that issues are identified as early as practicable, to reduce the time and expense associated with redesign.

The means of verification and validation for each requirement should be defined in the initial requirement. Although most of the activities described above can be performed analytically, it is often necessary to provide assurance through human-in-the-loop evaluations on prototypes, simulators or actual equipment. Because the scope of testing required to prove compliance with human factors requirements is not that well understood on many projects, it is important to establish, as part of the test plan, anything which significantly impacts this scope. For example:

- the number of scenarios against which suitable human performance would need to be demonstrated;

- whether or not it would be suitable to use a simulator or whether some tests would need to be conducted using actual equipment; and

- the number of team participants required, etc.

Methods, which can be derived from available best practice, and pre-requisites (e.g. suitable procedures, equipment and operators) should be established as part of normal project planning.

4. Lesson learning

Few would argue that an effective and widespread lesson learning process is essential to ensure that error management within safety critical systems is continuously informed and improved. The difficulty arises in knowing how to achieve it.

Effective lesson learning can be achieved through a variety of means. For example, establishing a risk register and keeping it active and updated with operational risk data provides a means of ensuring that everyone within the organisation is aware of the current risk situation and work that is ongoing to resolve specific risks.

Establishing focal points for specific risk reduction activities ensures ownership of key system risks and accountability for ensuring lesson learning is disseminated throughout the organisation. Ensuring regular reviews of the progress of risk reduction actions against safety targets is another way to ensure that the focus on reducing risks is not lost. Continuous performance monitoring, in particular the communication of that information throughout the organisation, is key to ensuring widespread and effective lesson learning.

Once system improvements go into service, the processes used in understanding current system risk come into play again to enable their ongoing performance in meeting the risk reduction targets in the operational environment to be assessed. For example, current operational risk data could be used to determine if a particular piece of equipment was successful in achieving specific risk reduction targets.

This process enables a number of useful questions to be asked:
- If the system improvement did not achieve the targets, why was it unsuccessful?
- If it exceeded them, were the targets set realistic?
- How can this inform the development of future targets for other system improvement projects?
- What action is required to manage safety risk?
- Who in the organisation needs to be aware or take action on this?

In order to ensure an informed organisation focussed on error management, lesson learning needs to extend beyond the operations room to a company-wide level. In particular, future systems designers, those working on safety cases, those making investment decisions about where to expend resource to best risk reduction effect and those procuring systems need to be covered by lesson learning processes. By ensuring salient lessons are learned throughout the company, the error management process can be better informed from inception right through the design lifecycle and into service. The overall aim is the design and implementation of more error tolerant and effective systems.

5. Conclusion

We know that the ability to manage the safety risks associated with human error is important for complex systems. Although there are difficulties associated with addressing all of the possible combinations of factors that impact human performance, this paper supports the view that systematic and informed consideration of human error at all stages in the system lifecycle can provide significant risk reduction benefit. To better inform decisions about human error risk, it is important to build our body of knowledge about human performance in a given complex system. This can be achieved by using a framework structured in terms of generic goals, functions, failure modes, causal factors and mitigations. Using a standard framework enables specific data to be shared and validated across all stages of the system lifecycle. In time, this will provide better information and will improve the integration of human error into the overall safety risk management process.

References

Chemical Industries Association CIA (1977). A Guide to Hazard and Operability Studies. Prepared by ICI. London: Chemical Industries Association.

Federal Aviation Administration (FAA). (1990). Profile of Operational Errors in the National Airspace System: Calendar Year 1988. Washington D.C.

Kinney, G.C., Spahn, J., and Amato, R.A. (1977). The Human Element in Air Traffic Control: Observations and Analyses of the Performance of Controllers and Supervisors in Providing ATC Separation Services: Report Number MTR-7655. McLean, VA: METRIEK Division of the MITRE Corporation.

Reason, J. (1990). Human Error. Cambridge: Cambridge University Press.

Shorrock, S.T. (1997). The Development and Evaluation of TRACEr: A Technique for the Retrospective Analysis of Cognitive Errors in Air Traffic Control. MSc (Eng) Thesis: The University of Birmingham: September 1997.

Shorrock, S.T., Kirwan, B., MacKendrick, H. & Kennedy, R. (2001). Assessing human error in air traffic management systems design: methodological issues. Le Travail Humain, 64, (3), 269-289.

Wickens, C. D. (1992). Engineering Psychology and Human Performance (2nd Edition): New York: Harper Collins.

Measuring & Managing Culturally Inspired Risk

Dr. Martin Neil[1]
Mr. Roger Shaw[1]
Mrs. Sheena Johnson[2]
Mr. Bob Malcolm[1]
Professor Ian Donald[2]
Dr. Cheng Qiu Xie[2]

Abstract

The search for improved service integrity, whether it is safety within hazardous industries or businesses seeking service probity, has thrown the spotlight on organisational culture. In some industries poor safety culture has been the root cause of publicly visible accidents, while poor operational culture has been the root cause of losses in the financial sector. This paper describes work undertaken on the EPSRC funded SCORE project to investigate how cultural issues can be incorporated into an organisations "risk account" through the use of Bayesian Networks.

1 Introduction

The SCORE (Sensing Changes in Operational Risk Exposure) project was set up in 2001 to investigate how the risk posed by poor organisational culture might be measured and monitored. This paper discusses culturally inspired risk and the problems of defining it. It will show how the project has defined and measured cultural risk and integrated that definition into software tools that permit organisations to perform "risk accounting". Two case studies will be discussed, one involving financial and operational risk and the other safety risk management. Within both of these cultural risk measurement plays a significant role.

Section 2 of this paper considers some of the background issues related to organisational culture and, in particular, why culture is now seen as central to the risk management equation. Bayesian Networks, the core technology used on the project, are then briefly described in Section 3. Sections 4 and 5 discuss two case studies undertaken by the SCORE project. The first relates to air traffic control and shows how a Bayesian Network was developed to model the likelihood of an air incident and how cultural concerns were incorporated into the model. The second case study focuses on operational risk measurement and again shows the manner in which cultural concerns were incorporated. Section 6 discusses the measurement of

---

[1] RADAR Group, Department of Computer Science, Queen Mary, University of London.
[2] Safety Research Unit, Psychology Department, Liverpool University.

cultural risk and Section 7 summarises the current state of the project and what work remains to be completed.

2 Background

Examples of safety culture, or rather the lack of safety culture, abound. The September 30th edition of Newsweek [Newsweek 2002] contained an article headed "Breach of Faith – A major nuclear-safety scandal raises troubling questions about Japan's culture of secrecy." In this article we are told that The Tokyo Electric Power Company (Tepeco), Japan's largest nuclear plant operator, admitted falsifying safety inspection reports for three of its fifteen reactors. These reactors, the report continues, had cracks, leaks and other safety breaches dating back to 1986. Further, Hitachi has also admitted that it withheld information on safety breaches in its equipment in order to keep nuclear facilities running. Then, in August this year, General Electric International, which built and maintains Tepeco's plant, admitted to falsifying safety records at thirty seven locations. In the light of these revelations Tepeco maintain that the facilities pose "no safety problems".

Japan produces around a third of its electricity requirements from nuclear powered generators and thus the country's economy is highly dependent on this source of energy. Further, like many nuclear generators, Tepeco is under a lot of financial pressure with debts of over $100 billion and likely losses of $1million per day for each generating station it shuts down for repairs. The Newsweek report goes on to say that the regulatory regime is lax with the power companies left to regulate themselves. In this regime they file reports that are signed off but rarely checked by Japan's Nuclear and Industrial Safety Regulator (NISA). As an indication of this lax regime, we are told, that in July 2000 an American inspector working for General Electric International found twenty nine discrepancies between his own inspection findings and those reported by Tepeco. The inspector is reported to have informed NISA who in turn told Tepeco that they had a whistle blower inside their ranks. Rather than act, NISA then allowed Tepeco to deal with the problems internally. Newsweek reports a Japanese Trade and Industry official as saying that "problems are negotiated and settled under the table. Danger is never a consideration". Further, in the light of the recent revelations about Tepeco, NISA has refused to prosecute the company and refuses to state why. Tepeco, following an internal investigation related to these latest revelations, has suspended five senior executives but has subsequently re-hired them as consultants.

Corporate ethics have recently come under the microscope as a result of various irregularities reported at companies such as ABB, Adelphia, Allied Irish Bank, Andersons, Elan, Enron, Global Crossing, Kmart, Merck, Qwest Communications, Worldcom, Xerox and others. All have resulted in allegations of dubious accounting practices, fraud and failures in internal auditing and risk management processes. However, the main losers have been shareholders and the root cause a failure in probity, corporate ethics and ultimately corporate-culture.

What then do we mean by corporate culture and what specifically do we mean by a failure in corporate culture? If we define corporate culture to be "the moral, social and behavioural norms of an organisation based on the beliefs, attitudes, priorities and actions of its members" then we are forced back to looking at the shaping factors of these beliefs, attitudes, priorities and actions. These are less easy to define because they ultimately depend on national norms, which in turn may derive from professional codes of practice, regulatory regimes and the extent to which these are policed. For example, in the case of Japan it would appear that NISA (the regulator) does lay down specific guidelines but these are not enforced as preference is given to "under the table agreements" which tend to be difficult to discover in a culture of secrecy. In a similar manner the investigations into the Baring's Bank collapse revealed failings both within the bank and on the part of its regulator, the Bank of England. Similarly, the recent corporate collapses in America have been attributed to weak accounting practices and the cosy relationships that have developed between clients and accountancy practices offering both consultancy and audit services to the same client. Don Cruickshank, chairman of the London Stock Exchange, has stated [Scotsman 2002] that "The probability of such a scandal happening here is less than in the US". He went on to say, "We have a stronger accounting profession in the UK, and have better corporate Governance than in the US, and the probability of Enron activity is less – though there's no guarantee that it won't happen." Having defended Corporate Governance rules he said that not all firms honoured both the spirit and letter of the law. "Probably in principle they are fine. Its whether in practice our major companies are honouring them in the spirit? Most do."

What we have attempted to indicate here is that corporate culture is a difficult concept to grapple with. It embodies the behaviour of organisations, the behaviour of those briefed to enforce standards whether they be safety standards, accounting rules, operational risk standards etc and it also encompasses the prevailing milieu within a society which will determine the degree to which laws and regulations are expected to be complied with in practice and the degree to which a free press is able to investigate potential scandals.

The aim of the SCORE project is to investigate how aspects of human error and culture, as they relate to the operation of complex systems (technical, financial etc.), influence risk exposure and the way in which it can be managed. In addition, and through the use of Bayesian Networks, the project is developing risk models that highlight the impact of these issues and make visible their effects in what is termed an organisations "risk account".

In the rest of the paper two case studies are described where we have modelled the impact of cultural risk in an air traffic control and a financial environment. In the first case cultural issues are examined in terms of their impact on safety risk and in the second on financial risk. However, we start with a brief description of the modelling approach we have adopted, namely Bayesian Networks.

3 Bayesian Networks

Bayesian Networks (also known as Bayesian Belief Networks, Causal Probabilistic Networks, Causal Nets, Graphical Probability Networks, Probabilistic Cause-Effect Models, and Probabilistic Influence Diagrams) provide decision-support for a wide range of problems involving uncertainty and probabilistic reasoning. The underlying theory of BNs is Bayesian probability theory and the idea of evidence propagation through a network structure. Although this theory has been around for a long time it is only in the last few years that efficient algorithms, and tools to implement them, have been developed to enable propagation in networks with a reasonable number of variables. The recent surge of interest in BNs is due to these developments, which mean that for the first time they can be used to solve realistically sized problems.

A BN is a directed graph, together with an associated set of probability tables called Node Probability Tables or NPTs. The graph consists of nodes and arcs as shown in Figure 1. The nodes represent variables, which can be discrete or continuous. For example, the node *Faults in Test/Review* is discrete having values 0,1,2,.. whereas the node *System Safety* might be continuous (such as the probability of failure on demand). The arcs represent causal/influential relationships between variables. For example, the *Correctness of Solution* and *Accuracy of Testing* influence the number of *Faults in Test/Review*; hence this relationship is modeled by drawing appropriate arcs as shown. The key feature of BNs is that they enable uncertainty to be modeled and reasoned about. BNs also offer considerable analytical power to the modeller. They can show which variables contribute most to establishing a hypothesis and can show how sensitive variable valuation is in determining a hypothesis. In more complicated networks variables may contribute in different proportions to establishing a hypothesis. For example, knowing how sensitive a variable is in determining a hypothesis will provide guidance on how much effort should be devoted to accurately determining the value of the variable. If small changes in the value of a variable dramatically alter the value of the hypothesis variable then it should be determined as accurately as possible; if the hypothesis is not that sensitive to the value of the variable then less effort can be devoted to its determination.

As part of the process of developing a BN, Node Probability Tables (NPTs) need to be constructed. Variables with no parents represent our prior assumptions regarding their likelihood while variables with parents will have probabilities conditioned upon those parent nodes. Determining the NPTs is part of the knowledge elicitation process, which traditionally has proved to be a time consuming and often difficult task. Although not discussed here the SCORE project has developed an approach to facilitate the collection of this information from experts and uses a small set of graphical tools to support the process.

The benefits of BNs may be summarised as follows:

- Provide perhaps the best method for reasoning under uncertainty.

- Permit the combining of diverse data, including subjective beliefs and empirical data.
- Predictions can still be secured even when evidence is incomplete.
- Permit powerful "what-if" analysis to test sensitivity of conclusions.
- Incorporate a visual reasoning tool, which aids documentation.

Experience in the safety and software domains shows that BNs can be used on real, large-scale risk problems.

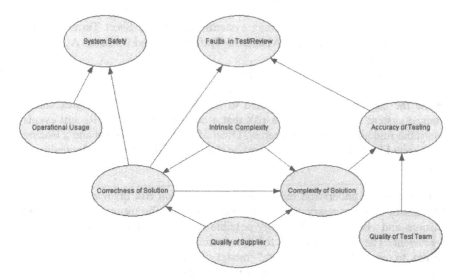

Figure 1: Example Bayesian Network

The use of BNs, as outlined above, is based on a decade old research programme, which started with the DTI/EPSRC project DATUM [Fenton *et al* 1998] in 1992. This project investigated how to integrate different sources of evidence together into a coherent and quantitative argument. The motivation was the production of what are called Safety Cases. Various approaches were investigated and BNs were chosen as the most suitable way of achieving this objective. With this decision work progressed on the EC funded SERENE project (1996-1999) [Fenton 1999] to develop a method for representing safety arguments using BNs. The results from this project included a fully developed method, a supporting tool for the method and a number of case studies. Since then numerous other EPSRC funded research projects, for example IMPRESS [Neil *et al* 1996] and SCULLY [Fenton Neil 2001] have contributed to the development of this technology. Various commercial applications have also been built using BNs including AID for Philips Research [Fenton *et al* 2002] and TRACS for QinetiQ [Neil *et al* 2001]. For those interested in reading further on the subject of Bayesian Networks reference should be made to [Pearl 1988], [Jensen 2001] and [Neil 2000].

4 Air Traffic Control

The purpose of an air traffic control (ATC) system is to control the trajectories of aircraft in such a way as to minimise the risk of collision. This objective is achieved through a range of measures that include:

- The design of the airspace (the virtual corridors within which the aircraft fly).
- Procedures for controlling access to, and use of, the airspace.
- The use of highly skilled and trained air traffic controllers.
- The use of collision warning systems such as the Short Term Conflict Avoidance system (STCA) and the Traffic Alert and Collision Avoidance System (TCAS).
- The skills of the pilots themselves.

Despite the care and attention devoted to the design of such a complex socio-technical system there is always a risk that a collision will occur. Recent events such as the mid air collision between a TU154 of Bashkirian Airlines and a Boeing 757 operated by DHL over Ueberlingen on the 1st July 2002 point to the reality of such risks.

In order to better understand the nature of the collision risk involved in an ATC environment it is necessary to have a model showing the means taken to avoid collisions and via this model identify possible factors that might arise and lead to an accident. The model adopted for this exercise is a barrier model, sometimes referred to as a "defence-in-depth" model. This model depicts a number of defences, which collectively aim to prevent accidents arising.

In order for a collision to occur various "events" must take place. The aircraft must have been on a collision course, the automated collision warning systems must have failed in some way, air traffic control staff must have failed to note the collision trajectory and rectify matters and the crew of the aircraft must have failed to detect and avoid the collision. Thus, when accidents do arise they can be seen as resulting from the breach of a number of defences.

Figure 2 shows a schematic of the barrier model taken as the starting point on this project. The first barrier is the design of the airspace so as to channel aircraft in such a way that they are kept apart, and to minimise the risks associated with crossing points where corridors meet (density and complexity). As well as serving safety objectives air space design has to serve economic and efficiency requirements as well.

Given a designed airspace, and given the demands of airlines and airports upon that airspace, flight planners have the task of assigning flight-paths to aircraft. They may take more or less cognisance of the risks arising from the combination of routes they assign.

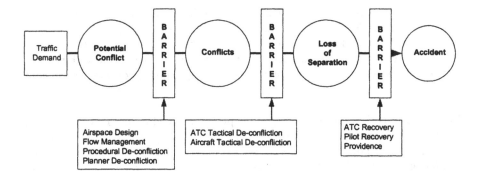

Figure 2: Schematic Barrier Model

Flow management is a function of air traffic control which, given warning of upcoming flight-plans, aims to maintain the volume of traffic though the various ATC sectors within manageable levels, so called "target sector flows".

In some sectors there is a distinction between "planning controllers" and "tactical controllers". While this is not true of all sectors, there is a useful distinction to be made between the two roles, even if performed by the same person. The planning controller is concerned with routing through the sector, while the tactical controller is concerned with the moment-by-moment management of aircraft, given that routing. Planners, given the anticipated flow of aircraft, attempt "planner de-confliction" by routing them through a sector in such a way as to minimise the difficulties faced by the tactical air traffic controllers - who have the job of maintaining separation of aircraft, normally through "procedural de-confliction" - i.e. following procedures designed to maintain separation.

After the tactical air traffic controllers, the next barrier is the Short Term Conflict Avoidance system (STCA), which, where installed, automatically detects impending loss of separation and raises a warning to the air traffic controllers to attempt 'tactical de-confliction'.

If, despite the barriers thus far, a potential conflict situation arises between two aircraft, then if they both have Secondary Surveillance Radar (SSR) transponders and at least one has a system called TCAS (Traffic Alert and Collision Avoidance System) installed, then TCAS should automatically provide recommendations to the pilot(s) for avoiding action(s) so as to achieve "aircraft de-confliction".

If an incident is still impending despite these barriers, then the situation is left for "last minute see and avoid action" by the pilot or providence (the trajectories are such that an accident is avoided).

The model described above presents a perspective on safety management. However, it does not represent any organisational issues. For example, airspace over a large geographical area tends to be broken into a number of sectors some of which will experience fairly light loading while other sectors will encompass areas of heavy traffic. Further, there are different types of sector, some concerned entirely with through flights and some with takeoffs and landings. Each sector has its own internal structure, depending upon traffic loadings, and within these internal structures teams of people will be found providing coverage throughout the day and night.

In order to provide a tractable exercise the project decided to focus its attention on a particular organisational model, which has sufficient individuation to allow cultural and psychological factors to be incorporated. Based on the barrier model and the organisational structure outlined above the SCORE project team have built a Bayesian Network model.

The BN, shown in Figure 3, captures various aspects of the barrier model described above. To the left of the network can be seen a series of nodes that represent the density and complexity of the managed airspace. The more complex the air space, the more aircraft in the airspace then the likelihood of an incident will be higher than for more lightly loaded and less complex airspace. Feeding along the top of the network is a calculation yielding the likelihood of an adverse event, condition by the volume of managed traffic. Planner performance, influenced by airspace volume and complexity, as well as staff competence and cultural factors, yields a plan. This plan, operated by the Tactical ATC staff, will influence the complexity of the tactical air traffic control task and will also yield a probability of an adverse event, conditioned by traffic volume. Penultimately, if the STCA is activated then the likelihood of an adverse event will be influenced by the competence of the Tactical ATC staff and the complexity of the encountered control situation. Finally, if loss of separation occurs, then the likelihood of a collision will be determined by the performance of the pilots in the face of TCAS activation.

Cultural influences are modelled in the lower portion of the net and Figure 4 expands that part of the network. Two distinct staff functions are shown, those performed by Planners and those by Tactical ATC staff. Each is subject to its own cultural influences as depicted by the nodes Culture Unit A and Culture Unit B. However, as both functions exist within a corporate culture, the Organisational Culture node reflects this. Thus the cultural influences on the two groups of staff come from both "work group" sources (in some cases strong and in other cases less so) and the overall organisation. The two nodes titled Cultural Dependence permit these two sources of influence to be accommodated allowing for a balance between the cultural sources to be struck or a preponderance of influence being given to one or other source.

Determining how to measure these two influences is still a matter of research, which is currently being pursued by the Safety Research Unit at Liverpool University and which is discussed further in Section 6.

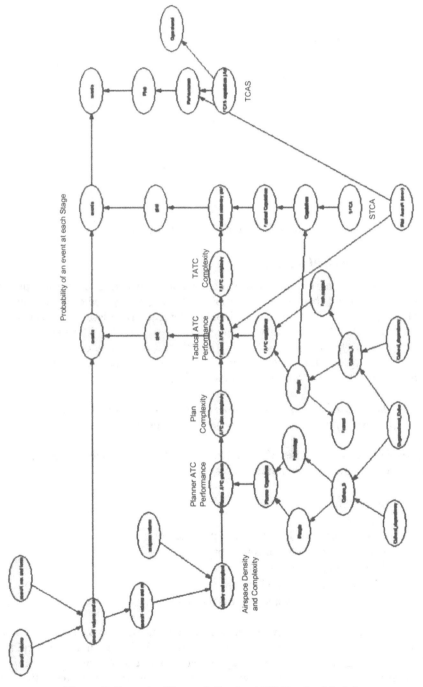

Figure 3: Bayesian Network for the ATC Barrier Model.

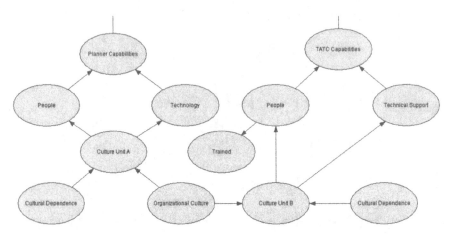

Figure 4: Cultural Nodes in ATC BN

5 Operational Risk

For financial institutions the concept of operational risk is currently a topic of growing importance, because of the emphasis the regulatory bodies are beginning to place upon it. The Bank of International Settlements (BIS) issued a consultation paper [BIS 2002] to identify "Sound Practices for the Management and Supervision of Operational Risk" which is the latest in a series of steps to ensure such risk is given the emphasis it deserves in a bank's risk management frameworks. This paper defines operational risk as:

> *"the risk of loss resulting from inadequate or failed internal processes, people and systems or from external events"*

It goes on to detail a number of operational event risks that could potentially result in substantial losses for institutions. These include:

- Internal fraud.
- External fraud.
- Employment practices and workplace safety.
- Clients, products and business practices.
- Damage to physical assets.
- Business disruption and system failures.
- Execution, delivery and process management.

Cultural issues are now being recognised as an important risk factor and it is acknowledged that management need to be aware of the important role that culture plays in handling risk. It is not enough to have excellent controls if they are not being adhered to by employees, which, again, is emphasised by the BIS [BIS 2002]:

> *"Both the board and senior management are responsible for creating an organisational culture that places a high priority on effective operational risk management and adherence to sound operating controls."*

Analysis of high profile loss events such as the collapse of Barings Bank and the recent Allied Irish Bank (AIB) loss serve to illustrate the importance of cultural issues. Analysis of the AIB case [Wachtell et al 2002] reveals the role that cultural factors play in this event and also how the monitoring of these types of issues can result in a stronger control environment than would arise from assessing which procedures and systems are in place.

The losses incurred by AIB were related to fraudulent activities uncovered in February 2002 in the trading operations division of its US subsidiary, Allfirst, with losses described at £473 million. The subsequent report highlighted the weak control environment and inadequate monitoring with specific instances being described in more detail, a number of which are outlined below.

- Supervision was described as inconsistent and unreliable due to the physical location of offices.
- There was a potential conflict of interests with the same person being in charge of both profits and controls.
- Confirmation of options was not obtained by back office staff even though controls stated they should be.
- Checks on figures were not completed correctly (by employees, internal audit and treasury)
- Rusnak, (the person accused of fraud) was described as having a bad temper and behaving in a bullying fashion. For example he threatened to leave if back office staff continued to question his actions. Further, his supervisor added, Rusnak's departure would lead to subsequent job losses.
- Rusnak took advantage of employees who facilitated him in circumventing controls through inexperience, poor training, poor supervision and in some cases laziness.
- There was ambiguity in the workplace as to who should be supervising particular activities.
- Risk reporting practices needed to be more robust.

It is clear from this very brief outline of the full report that in addition to poor controls there were also instances where controls, already in place, were not being adhered to. Although at times staff members queried these lapses, management took no action. Indeed, the report comments on this:

> *"the failure by treasury management to follow through on back office inquiries may have contributed to an attitude among*

operations staffers that the confirmation process was a pointless formality."

Analysing the risk culture of an organisation will result in the early identification of issues, which may reduce the likelihood of certain types of risk. Knowledge of any weaknesses in the control environment would give an organisation the opportunity to act upon weaknesses before a loss has occurred, or to recognise and curb losses at an earlier stage. Listed below are some of the points a risk culture analysis would investigate:

- Employee's evaluation of risk controls in the workplace.
- Personal working practices.
- Risky working practices.
- Co-worker's perceived involvement in risky working.
- Workforce's perceived risk encouragement and support.
- Co-worker's perceived involvement and evaluation of the control.
- Management's perceived evaluation of the control system.
- Participative communication.

The second SCORE case study is addressing these specific issues and by way of illustrating the work currently being carried out a model will be described which shows how risk factors associated with the strong or week promulgation of risk management practices between parts of an organisation can be highlighted in a Bayesian Network.

Because the BN of the Operational Risk model is too large to show here a schematic will be used. Figure 5 shows how operational risk can be modeled.

The left hand side involves assessing the capability of an organization's security controls. The effectiveness of these controls will be determined by a number of factors, namely resources and budgets, staff competence, management competence, technology and operating procedures. These in turn are heavily influenced by the risk culture of the organization. An audit of the security controls will produce an assessment of their effectiveness, but the trust you put in this audit result will depend on whether the audit is accurate. This will, in part, also depend on organizational culture. The top right hand side of Figure 5 shows a number of threats to the organization. These include internal and external fraud, reliability of technology systems and vulnerability to disasters, for example, terrorist threat. These are combined, and the influence of the risks they present is reduced by the controls process capability. Finally, vulnerability is used to forecast the number of operational losses the organization might expect to experience. As actual losses are observed, these can be used to update our estimate of vulnerability; however, the extent to which the loss data can be relied upon depends on how good the loss reporting and data collection system is. In poorly managed institutions, financial losses can remain hidden for substantial periods of time; they may accrue on a daily basis and be realized only upon discovery.

Figure 6 shows a segment from the actual BN model with some of the marginal probability distributions displayed. For example, the Culture node (top left) shows that informed opinion believes that around 10% of the organisation's business units have a bad culture and 90% have a good culture. Similarly, 27% have poor Operational Procedures and 73% have good procedures. Determining these probabilities is part of the elicitation exercise needed to populate the BN's probability tables.

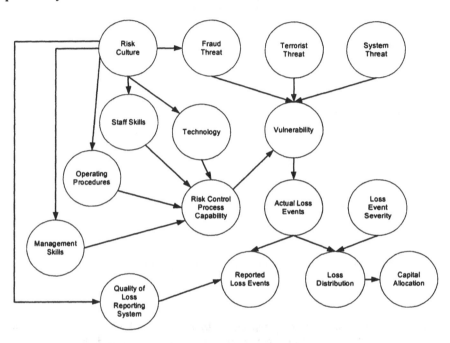

Figure 5: Schematic Operational Risk BN Model

Figure 7 reveals what happens when we enter some observations relating to one of the organisation's business units. Assume that we assess that this business unit has poor Operating Procedures and that the Management Skills are also poor. Further, we know that the average size of business transactions is in the region of £10,000 and that there has been one reported loss. Entering and propagating this information we see that the Culture node has changed dramatically suggesting a poor culture. Further, the Vulnerability distribution has changed indicating the current perceived risk and the Predicted Loss node has changed indicating a revised loss distribution, in this case upwards. Based on these observations an independent audit of this business unit would probably be justified

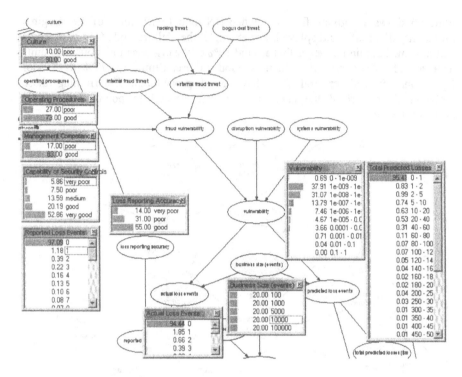

Figure 6: Marginal Probabilities for the Operational Risk Model

Figure 8 takes an alternative perspective. In this case all we know is that our measurement of Culture (see Section 6 below) has determined that it is poor. Further, we know that compliance with Operating Procedures is poor and Management Skills are poor. Entering and propagating this evidence and then comparing the resultant probabilities with those shown in Figure 6 reveals a significant rise in Vulnerability and an increase in the estimate of Potential Loss. What is significant about the change in Potential Loss is the growth in the tail of the distribution, which clearly illustrates the potential for very high severity events.

If this analysis were undertaken for each of our organisation's business units then combining the predicted loss graphs would yield an overall predicted loss distribution for the organisation as a whole.

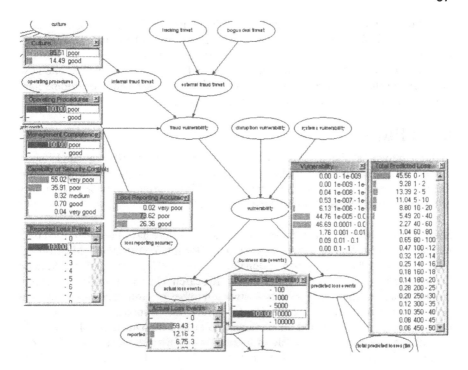

Figure 7: Entering and Propagating Observations

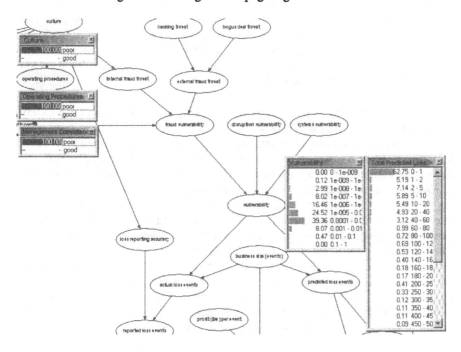

Figure 8: The Effects of Culture on Vulnerability and Potential Loss

A note of caution should be sounded. The effects shown in Figure 6, Figure 7 and Figure 8 result from the node probabilities that are part of the net. Determining these probabilities is not trivial and research, as described in Section 6, is currently underway to decide how to model cultural risk within these net structures and how to determine the associated node probability tables, in other words determine how to model the effects of culture on risk taking behaviour.

6 Culture

In both networks shown above culture has appeared as either a single node representing organisational (risk) culture or in several nodes representing more localised cultural influences originating from different groups of people (operators or business units). What do these nodes represent?

Safety culture and the factors underpinning safety culture have been identified by a number of researchers in a variety of settings [Donald, Young 1996] [Mearns et al 1998], [Lee 1998]; see also [Donald, Shaw 2002]. The Safety Attitude Questionnaire (SAQ) used by the Safety Research Unit is a self-report safety culture measure where respondents answer a number of attitudinal questions on a seven-point Likert scale. In order to ensure the responses are honest the questionnaires are completed anonymously to ensure individual people cannot be identified and a true safety profile of a group is established. Analysis of the SAQ enables a risk profile incorporating a number of factors to be drawn up at a number of levels, for example the overall safety culture of the organisation or the safety culture of distinct groups within the organisation. These groupings can be quite varied in nature, for example departments, tenure, shift patterns worked etc. The profiles can then be used to establish any areas of the organisation reporting a weak safety (risk) culture and which may therefore be at increased risk of an incident occurring. For further information on the analysis of safety culture using the SAQ see [Donald, Young 1996], [Donald, Canter 1994].

This approach had been used in a number of working environments and in a variety of different countries and the SAQ has proven to be a reliable and valid instrument in measuring safety culture in industrial settings. The current research is focusing upon using the instrument in other working environments, for example ATC and financial organisations. Any instrument needs to be applicable to the environment in which it is being utilised and therefore any changes in terminology and emphasis are made to the SAQ prior to the measure being used. This is achieved through interviews and focus groups with people working within the organisations. Where necessary, additional items are included to ensure all relevant points are covered. The ATC environment, for example, retains the focus upon safety and therefore the measure will retain most of the original terminology and content. However, when the focus is upon financial risk rather than safety the measure is altered to ensure the terminology used is appropriate for that sector. This is in relation to both the wording of individual items and also the outcomes, for example accidents/near misses in safety and incidences/financial loss in finance.

Although the SAQ is fine-tuned to ensure its application within specific sectors it is expected that the **underlying factors** will be the same. That is, just as there are risky working practices that influence workers involvement in accidents so too will there be risky working practices that influence financial loss events. It is likely then, that the same organisational, social and psychological processes are involved in risk to safety and financial loss.

Once the risk culture of these organisations is identified the next step is to build the culture measures into the Bayesian Network to enable the cultural influences of risk to be analysed in relation to other risk elements of the organisation. Both the risk culture of the organisation as a whole and the risk culture of any relevant groups within the organisation will be represented within the BN. In addition to this the individual factors that are most predictive of involvement in incidences will also be present. This ensures that the most predictive factors in relation to a specific sector can be linked to other outcomes within the organisation. For example, previous research by the Safety Research Unit into safety culture demonstrates that the three most powerful factors that discriminate between people who have been involved in an accident and those that have not are attitudes towards risky working practices, the employee's personal evaluation of the safety system and the workforces perceived evaluation and involvement in safety meetings. It is expected that salient factors such as these will be present in the net, linking in to both the overall risk culture node of the organisation and also to specific outcomes. For the financial risk culture research it may be that other factors are more predictive than the safety factors outlined previously, these will of course be identified through the risk culture research and subsequently represented within the risk culture net.

Once the risk culture of an organisation is established comparisons can be made both within the organisation itself, that is between groups or departments, and to other organisations working within the same sector. A normative database already exists for safety culture, which enables organisations to compare their safety profile for benchmarking purposes. The ongoing research into financial risk will enable a normative database of risk culture in the financial industry to be compiled. In addition to this, further comparison can be made within organisations if their risk culture is measured on separate occasions, which enables changes in the risk profile of organisations to be identified and any impact this may have on the outcomes the organisation experiences analysed. In the context of the Bayesian Network any changes within the culture of an organisation can be entered into the net and the subsequent influence this may exert on other risk elements within the net, including outcomes can be established. Section 5 gives examples of this type of net investigation.

7 Ongoing Work and Conclusions

The work described here is well advanced. In the case of the ATC example existing data and expert evidence is being collected to populate the node probability tables. Data for the cultural nodes will be derived from existing work carried out by the SRU at Liverpool University although these results are currently being validated for the environments within which SCORE is focusing. Once this work is completed the operation of the model will be investigated in terms of historical incident data. Further, the ability of the model to suggest changes in risk exposure arising from changes in the cultural variables will also be examined. Finally, attention is being given to the elicitation process in order to develop procedures that help experts determine the probabilities needed within the networks.

References

[BIS 2002] Bank for International Settlements (2002) Basel Committee on Banking Supervision: Sound Practices for the Management and Supervision of Operational Risk July 2002

[Donald, Canter 1994] Donald I, Cantor D. Employee Attitudes and Safety in the Chemical Industry. *Journal of Loss Prevention in the Process Industries, 7*, 203-208, 1994.

[Donald, Young 1996] Donald I, Young S. Managing safety: an attitudinal-based approach to improving safety in organizations. *Leadership and Organization Development Journal, 17*, (4), 13-20, 1996.

[Donald, Shaw 2002] Donald I, Shaw R. *Safety Culture* in Safety Systems (The Safety-Critical Systems Club Newsletter) May 2002. Volume 11. No 3.

[Fenton *et al* 1998] Fenton N, Littlewood B, Neil M, Strigini L, Sutcliffe A, Wright D. Assessing Dependability of Safety Critical Systems using Diverse Evidence, IEE Proceedings Software Engineering, 145(1), 35-39, 1998.

[Fenton 1999] SERENE Consortium, "SERENE (SafEty and Risk Evaluation using Bayesian Nets): Method Manual", ESPRIT Project 22187, 1999. http://www.dcs.qmw.ac.uk/~norman /serene.htm.

[Fenton, Neil 2001] Fenton N and Neil M, Making Decisions: Using Bayesian Nets and MCDA, Knowledge-Based Systems 14, 307-325, 2001.

[Fenton *et al* 2002] Fenton N, Krause P, Neil M. Software Measurement: Uncertain and Causal Modelling. IEEE Software. July/August 2002.

[Jensen 2001] Jensen F. Bayesian Networks and Decision Graphs. Springer-Verlag. 2001.

[Lee 1998] Assessment of safety culture at a nuclear reprocessing plant Work and Stress, 12 (3), 217-237, 1998.

[Mearns *et al* 1998] Mearns K, Gordon R, Flemming M. Measuring safety climate on offshore installations Work and Stress, 12 (3), 238-254, 1998.

[Neil *et al* 1996] Neil M, Littlewood B, and Fenton N. "Applying Bayesian Belief Networks to Systems Dependability Assessment". Proceedings of Safety Critical Systems Club Symposium, Leeds, Published by Springer-Verlag, 1996.

[Neil 2000] Neil M, Fenton N. Building Large Scale Bayesian Networks. Knowledge Engineering Review. 15(3), 257-284, Sept 2000.

[Neil *et al* 2001] Neil M, Fenton N, Forey S and Harris R, "Using Bayesian Belief Networks to Predict the Reliability of Military Vehicles", IEE Computing and Control Engineering J 12(1), 11-20, 2001.

[Newsweek 2002] Newsweek, 30th September 2002.

[Pearl 1988] Pearl J. Probabilistic Reasoning in Intelligent Systems: Networks of Plausible Inference. Morgan Kaufmann, San Francisco, 1988.

[Scotsman 2002] The Scotsman, Thursday 11th July 2002.

[Wachtell et al 2002] Wachtell, Lipton, Rosen, Katz & Promontory Financial Group (2002) Report to the Boards of Allied Irish Banks, p.l.c., Allfirst Financial Inc. and Allfirst Bank Concerning Currency Trading Losses March 12, 2.

Safe Systems: Construction, Destruction, and Deconstruction

JM Armstrong
Senior Research Associate,
Centre for Software Reliability, University of Newcastle Upon Tyne,
Newcastle Upon Tyne, United Kingdom

SE Paynter
Senior Principal Engineer,
MBDA UK Ltd,
Filton, Bristol, United Kingdom

1   Introduction

Deconstructive Evaluation of Risk In Dependability Arguments and Safety Cases (DERIDASC) is a study focussed on the language used by safety engineers in their intellectual discourse. The DERIDASC project is inter-disciplinary in the sense that it experiments with techniques from philosophy, literary theory (Eagleton 1996, Culler 1997) and semiotics (Barthes 1994, Cobley 2001, Culler 2001) to diagnose problems of language, definition, and interpretation in safety engineering. The project aims to make safety engineers re-think and improve some of their habitual definitions. The project adopts methods of textual analysis usually found only in studies of the arts and literature, although the kinds of textual studies we propose have also been influential in the discipline of law (Ward 1998, Chapter 7).

In particular, the project is applying the ideas of "deconstruction" to safety texts. Deconstruction is a term coined by philosopher Jacques Derrida (see Abrams 1999 for a short summary) and denotes the analysis of a text to reveal hidden meanings, especially those which contradict the surface message. The critical reader reads a text "against the grain", concentrating less on what the author is trying to say than on issues such as what the text tries to avoid saying (e.g. the playing down of facts that might undermine what is argued) and on what is asserted rhetorically without evidence. The idea of deconstruction is to challenge the unconscious presuppositions inherited from conceptual frameworks for thinking. Deconstruction, it can be argued, facilitates the evaluation of the text and its language as 'technologies' for thinking (Clark 2000). We hypothesize that Derrida's deconstruction, with its emphasis on revealing suppressed contradictions and paradoxes in texts, may help illuminate some of the difficulties of safety argumentation and might be a basis for new analysis techniques.

In Section 2 of the paper we offer a brief overview of the problems of textual analysis as they are represented in Derrida's work and in deconstructive literary criticism in general. Our emphasis is on how the necessarily limited scope of a text renders it incomplete entailing risks of circular justification and self-contradiction. Section 3 explains 'deconstruction' as a procedure (although a 'deconstruction' should avoid reduction to a systematic 'procedure') for the analysis of binary distinctions. It presents an example deconstruction of the distinction between *likelihood* and *severity* that is commonly assumed in standard definitions of risk (see van der Meulen 2000, p. 245 for a list of examples). In Section 4, we discuss why some of the issues raised by deconstruction are necessarily of concern in safety engineering, and indeed are already having their effects.

2 The Problem of the Text

A text is a sequential structure requiring a starting point and a conclusion. It is essentially narrative in form. Thus we require a starting point for our explanation of deconstruction. We start with a discussion of some familiar problems of authorship; but by the very act, we have already run the risk of misrepresentation. This problem of choosing a starting point for a text, of laying out the basic assumptions of an argument, or of conceiving the axioms and definitions of a formal system, are key concerns of Derrida's deconstruction.

Most authors have experienced uncertainty about how to begin a text. A well-known cliché that purports to explain the usual structure (introduction-body-conclusion) is that it should "tell the reader what you're going to tell them, tell them, then tell them what you've told them". It seems to indicate that one should begin with the desired conclusion; but if this advice were to be followed rigorously a text would contain three attempts at assertion and no argument (even then, according to deconstructive thought, each attempt to repeat 'the same' meaning would be doomed to failure).

To be accepted as an argument as opposed to a series of rhetorical assertions a text needs starting points that appear in some way more basic than its conclusions. The choice of these starting points is absolutely necessary (no text could recount everything) but it imposes bounds on what can be said in the text. Thus each genre of text (a technical manual, an essay, a novel, a poem) depends upon a reader's acceptance of conventional assumptions about 'relevance' for that genre, and additionally, any particular restrictions requested by the author of the text at hand. There is a tacit agreement that certain subjects are to be considered 'irrelevant' to that genre of discourse, and will not be raised in it. This is an entirely necessary procedure, but unfortunately it is hard to argue that 'relevance' defines anything other than a socially-constructed boundary which the reality of events need not respect.

The 'starting point' (Derrida's term is 'origin') for any text or argument cannot be absolutely justified, since one knows that reality exceeds the bound it constitutes: before the argument begins, everything else has already 'begun' (Bennington & Derrida 1993). So when we consider what could determine a choice of origin(s) (of basic assumptions, or of founding axioms and definitions) it becomes apparent that the most attractive criterion is that they allow the desired conclusion to be reached by some means. Thus we encounter the problem of 'constructivism'. Origins of all kinds can be challenged on the basis that they are constructed in order to guarantee and preserve a desired conclusion, giving it the appearance of something that has been 'justified'; indeed, the basic assumptions may even be 'reverse engineered' from the desired conclusion.

Although it is possible for an author to start with some unusual or important observation and work through its possible consequences with no clear end in view, this is unlikely to lead to a highly structured and disciplined text (although literary authors have experimented with the approach). Nonetheless, most texts are constructed so as to mimic this process and in a successful text the conclusions will appear to follow naturally and obviously from the assumptions. However, this structure rarely corresponds to the process by which the text was really constructed especially when the writing of the text *demands* at least some idea of the desired conclusion. Even a text that sets out to answer a question in as unbiased a fashion as possible will tend to presuppose a fixed set of expected (or already extant) answers. Indeed, to ask a question in a text is already to have made a proposition regarding the significance of the question and the possibility of an answer.

One might object that in rational argument, which proceeds according to the laws of logic and of grammar, the process by which the text has been constructed does not matter. So long as the conclusions follow from the assumptions according to the laws and so long as the assumptions are accepted, the conclusion must be accepted. However, this is also true of a tautology, which raises the issue of how to distinguish tautological argumentation from 'meaningful' argumentation. For example, logicians have long realised that formal argumentation and proof cannot eliminate the possibility of circular reasoning. This is partly because no logic can prove that its own axioms and definitions are indubitable truths; but it is also because circular reasoning is not a problem of logic but a problem of belief (Cambridge 1999, p. 144).

We often accept a set of assumptions in a provisional sense 'for the purposes of argument', in order to initiate communication, or just to get through the task of reading a text; but why should we ever be more committal and accept assumptions as true thereby accepting the conclusion? Ironically, one possible reason for accepting assumptions is that the conclusion that follows from them is already believed, or at least strongly desired. Thus even the most rigorous argument cannot eliminate the risk that its assumptions have become 'detached from' the reality they are supposed to describe. Biases, perceptual limitations, and the desires of the arguer are important factors; but the necessity of structuring reality into a tractable mental model is the real culprit. Indeed, deconstructive thinkers argue that the very nature of the 'signs' we use to represent reality and their irreducible role in perception and thinking ensure that abstract arguments can never be unproblematically 'attached to' reality.

Deconstruction sees the relation between language and reality as a problem of 'effects of meaning' rather than one of logic. Questioning the meaning of assumptions is of course essential to any assessment of their validity. However, in *Limited Inc* Derrida shows that any attempt at an objective assessment of 'meaning' is bound to encounter highly inconvenient obstacles (Derrida 1977). When we consider 'effects of meaning' we find that they are inherently unpredictable. They are partly determined by socially accepted rules for communication, and partly determined by the individual intentions of the communicators; but they also depend upon what is rather vaguely referred to as 'context'. To complete the definition of a meaning effect, we would need to capture 'context' in a description; but context appears to have no definite bound; and furthermore, its description merely produces more 'text' which requires context to explain it in turn. One can try to arrest the indeterminacy of context by appeals to "what a speaker (author) must have meant at the time"; but this is another disguised appeal to context. It is complicated, for example, by the question of how unconscious mind relates to conscious intentions, and whether speakers (authors) really had clear intentions *in the context* of utterance (writing). Because of the apparent necessity and simultaneous impossibility of an objective definition of a context, nothing is more than provisionally sufficient to bound the effects of meaning that a particular text can have; these are, after all, a function of the future contexts in which the text will be repeated (read). There is therefore no 'final interpretation' of any text, even the most simple of sentences. At best we can expect meanings to be socially constructed and agreed; but this is hardly an objective foundation for rational thought.

It is common to represent the problems of natural language meaning in terms of 'ambiguity'. Derrida (1981) argues that this actually underestimates the extent of the difficulty, since it presupposes that a text has a fixed set of meanings that can be identified in principle, even if no objective choice can be made between them; but reading a text is always 'productive'. Because context itself is unpredictable and because we are condemned to shift through time, any text retains the potential for effects of meaning that cannot be foreseen. Each time we re-read a text we may notice (produce?) something new. As Derrida puts it, "iterability alters" (Derrida 1977, p.62); and it does so because previous contexts can never be perfectly described and are never exactly reproduced.

Wheeler (2000) summarises Derrida's thought as a working out of the consequences of there being no 'magic language'. A 'magic language' would be one to which all other languages could refer for their 'literal meaning'. Usually, we think of 'thought' or 'perception' as the things to which our utterances refer; but Derrida argues that these too involve the manipulation of signs, and are therefore linguistic, with all the interpretive insecurities and indeterminacies that natural language exhibits.

The problematic nature of the idea of a 'magic language' and of the opposition between 'literal' and 'figurative' meaning may not be immediately apparent; but it suffices to consider whether the 'literal meaning' of a statement could ever be written down non-figuratively in some language. If such a language existed we would be well advised to discard all others. Its meaning would never be in doubt, and it would provide an absolutely secure basis for thought and expression. All of poetry with its complex metaphors would be expressible 'literally' in it. The very fact that we can translate texts from one language to another seems to suggest that there is indeed some 'literal' meaning being transferred; but Derrida counters that the process of translation is always somewhat imperfect with respect to the source text and creates new meaning effects in the target text.

Since the late 1960's, Derrida has published around twenty major works arguing that philosophy has persistently resisted close consideration of the unpredictabilities of writing, of representation, and of interpretation because of the disconcerting implications for thought, rationality, and logic. If meaning effects are essentially unpredictable, we cannot be sure that we are thinking 'rationally' at all. He argues that the idea that meaning can be mastered, described, and bounded is an unreasoning prejudice which Western modes of thought secretly depend upon. He terms this prejudice 'logocentricity'. For Derrida, since a text is 'founded' insecurely on its meaning effects it is a rather perilous structure, condemned to conceal knowledge of its internal contradictions whilst also being condemned to reveal them to readers who will not play the author's game. Since it is in the nature of any 'sign' to be arbitrary with respect to its referent, there is only a conventional, not a natural, attachment between them. So when considered as a network of cross-referring signs, a text always has the potential to mean something other than what the author intended and perhaps even to contain a *denial* of what the author was trying to say.

Derrida's deconstructive analysis of the dilemmas of the text can be loosely compared to Godel's discoveries about incompleteness in formal logic. A text can attain an appearance of consistency only by self-imposed bounds on its completeness; as the text attempts a more complete treatment of its subject, internal inconsistency, loss of structure, and lack of conclusiveness (Derrida uses the term 'closure') become more apparent. A deconstructive reading concentrates on the way that terminology and criteria of relevance are 'constructed' in order to try and prevent a loss of control over meaning, whilst maintaining the appearance of an ordered progression from assumptions to conclusions. Thus instead of refuting the arguments in a text, a deconstruction concentrates on revealing the limits of authorial control. The challenge is that an apparently coherent, logical, and structured text achieves its ordered appearance not by representing truths, but by suppressing anything that contradicts or undermines its intended message or else renders that message trivial in its own terms. Deconstructive readings often seize upon issues mentioned only in passing in a text and attempt to show that when the implications of these issues are followed through rigorously they contradict the main line of argument.

The alleged inevitability of this kind of internal collapse derives from the fact that a text needs starting points - 'origins'. Deconstructive thinkers argue that since a starting point is a necessary imperfection, it is inevitable that in the course of thinking and writing ideas that cannot be properly accounted for in terms of the argument will crop up. How can a text deal with such ideas? It might resort to a kind of 'myth of The Fall' in which the valued origin is somehow tainted from without; it may fall silent at what appears to be a crucial point; it may try to evade the difficulty by subtle appeals to the readers background knowledge, culture, or charity; or it may resort to blatant rhetoric in order to divert attention from inconvenient facts, contrary views, or deep uncertainties; misrepresentation or caricature might be used in order to weaken the force of a contrary idea or opinion. Rhetoric is commonly used to assert that certain issues are irrelevant to an argument, which raises the question of how they came to be raised there in the first place.

All these phenomena are indicative of problems with the orderly progression of argument in the text. Indeed, the argument may not be 'progressing' at all; perhaps the desired conclusion is merely trying to re-assert itself in different terms, and since "iterability alters', inevitably failing to do so.

More positively, Derrida is interested in how, at times when argumentation and terminology falter, valuable progress can be made by a resort to analogy and metaphor. In rational discourse the status of figurative language is problematic; metaphor is regarded as an unsatisfactory detour on the way to truth; but deconstructive thinkers are generally positive about figurative language, and adopt Nietschze's dictum that 'truths' are merely metaphors whose status as such we have forgotten out of habit (for a discussion see Gayatri Spivak's introduction to Derrida's *Of Grammatology* – Derrida 1967). Another a-logical but quite productive strategy is the introduction of a neologism. Neologisms usually have recognisable roots in familiar words; indeed an author may reuse a familiar word without alteration, thus extending its meaning into new contexts. Derrida calls the study of this process 'paleonymics'. These processes produce new meaning effects whilst drawing on the power of existing words. Such strategies are not risk free. Neither are they obviously justifiable by any criteria except practical necessity. Yet without them argumentation would be exceedingly difficult, and perhaps even sterile or impossible.

3 Deconstructing Risk

The usual starting point for a deconstructive reading is the analysis of the binary distinctions that a text depends upon. Deconstructive thinkers claim that each term of a binary opposition will 'contaminate' the other. This view arises from a theory of language according to which the meaning of a term is determined by its position within the linguistic system, and not by any fixed property of 'meaning' that is indissociably bound to it. A 'meaning' is an effect produced by the inter-relationships among the terms of a language. Consequently, neither concept in an opposition of contrast has an identity that is entirely independent of its 'opposite'. For example, if we take one of the terms of an opposition and try to define it, we find that we can only do so by mentioning the other term, and vice versa. Each term contains what Derrida calls the "trace" of its opposite: so in deconstructive thinking, concepts are impure, or to put it another way, distinctions can always be undone by abstracting something that is common to both of them. Derrida claims that the usual result is an explosion of complexity of discriminations, rather than mere vagueness (Derrida 1977).

As an example we will examine the distinction between the *likelihood* (probability) of an accident and the *severity* of an accident. These two terms are broadly accepted as independent variables which combine to make up what we call 'risk'(van der Meulen 2000, p. 245). It might not be thought that either of these terms could be seen as more 'original' or 'valuable' than the other. Yet experts in probability theory tend to argue that however severe the consequences associated with a risk, if the probability can be made so small that the overall product (the total 'risk') is insignificant, then risk acceptance will be rational. Those who distrust probability theory argue that probability models are prone to give us the answers that we would prefer to hear, so that the severity of the possible consequences should be the primary criteria in risk acceptance. We can see this debate in terms of a disagreement about which component of risk should be of prime importance, even though the definition of risk as their 'product' seems to suggest that neither is primary.

The next deconstructive question will be to consider how severity and probability could be interdependent. They are presented as independent variables, but a close questioning of this assumption quickly complicates it. A measured probability of an accident, against which any estimate of likelihood is ultimately assessed, cannot be determined until the system lifetime has passed. However, this 'objective' measure can itself be determined by the severity of accidents. For example, suppose that an airliner crashes on its first passenger flight. Consider two possible futures: a) political pressure keeps the aircraft flying and it goes on to build up an enormous number of accident-free flight hours; b) the crash is used as a reason for cancelling the aircraft programme. The different lifetimes make the "objective" probability of an accident per flight hour very different in each case; but the aircraft lifetime is dependent upon the political will to either keep flying the aircraft or withdraw it. Concorde provides an example of this. A recent crash transformed Concorde from one of the safest aeroplanes in terms of accidents-per-flying-hour into one of the most dangerous (Daily Telegraph 2000); but the political will existed to keep Concorde flying. If it had not, Concorde would have remained a statistically 'dangerous' aircraft. It is possible that Concorde will go on in future to build up further flight hours without incident, thus apparently 'becoming safer'. The idea that a measured probability is 'objective' is undermined when we realise that the size of the sample space from which the measurements were taken is determined by something that is not an objective given. As far as statistical measures of accidents-per-unit-time are concerned, the size of the sample space is determined by willingness to carry on living with the severity of any actual consequences. Thus, not only are probability *estimates* dependent upon severity estimates - in the sense that we may unconsciously underestimate the probability of events we fear – but even the *measured* probability of an accident-per-unit-time is dependent upon its severity should at least one accident actually occur.

Further analysis also produces the explosion of complexity of meaning predicted by Derrida. We have argued that what one might think of as "objective" measures of risk *likelihood* depend upon subjective willingness to live with risk *severity*. However, we must further ask what determines our willingness (or unwillingness) to live with the possibility of severe consequences and why we might continue an activity even after it has already had severe consequences. Evidently, this question leads us to a consideration of enormously complex (even imponderable) issues of choice. For example, often in the aftermath of an accident we are led to reiterate the original question of risk acceptance. This may happen even where the accident falls within the expected probability of occurrence. If, after a (re)consideration of the likelihood of recurrence and the severity of the consequences, we make changes to the system to reduce future risk, we cannot avoid the challenge that we have retrospectively invalidated the original risk acceptance argument and that any new argument will be just as fragile as the first one. The rather-too-neat distinction between *likelihood* and *severity* perhaps (mis)represents risk as a simpler, more empirically verifiable, and more manageable concept than it really is. Yet it seems impossible to do without this distinction. Indeed, the 'deconstructive argument' given above itself depends upon it. Culler (1998, p.149) captures this paradox in an amusing way, noting that deconstruction can be described as "sawing off the branch upon which one is sitting".

4 Relevance To Safety Engineering

Derrida denies that deconstruction can be reduced to a corrective procedure or method, preferring to define the unravelling of conventional meanings merely as "what happens". Therefore, we do not propose the deconstructive analysis of texts as a replacement for the more usual procedure of analysing safety arguments for flaws in the reasoning, inaccurate data, and gaps in the evidence; but neither do we consider them as necessarily 'secondary' in relation to such tasks. Were we to do so, we would fall prey to a form of contradictory logic that Derrida (1967) calls 'the logic of the supplement'; for example, if we state that the textual analyses we propose are mere 'supplements' to current procedures, then we imply that they are both necessary *and* unnecessary to current procedures. Anything that can be 'supplemented' must by definition have a basic deficiency or lack that the supplement can remedy; there is therefore no reason to consider it as 'primary' or self-sufficient in the absence of its supplement. Deconstruction must already be going on in safety engineering, but not under that name. In this section we argue that this is indeed so.

The problems of pinning-down intended meaning, of finding implicit assumptions and circularities, and identifying what is a-logical, or even incoherent in a text are of interest to anyone involved in assessing safety texts. The text, with all its risks, is a technology that safety engineering seems unable to do without. Indeed, we are not the first safety experts to take an interest in deconstruction. For example, Turner (1994) discusses Derrida's ideas about the relation between chance and necessity, and how it relates to the interpretation of accidents. On DERIDASC we have turned to the textual representation of rational argument and rational choice in the process of safety acceptance.

For texts that present safety arguments the conclusion in view is specified in advance and necessarily so: naturally, the purpose of a safety argument is to arrive at the conclusion that a system is adequately safe. Once the argument has been constructed (usually by the supplier of a system) so as to reach this pre-specified conclusion, it is analysed for flaws according to whatever regulatory or assessment criteria happen to be in place in the application domain. The rules of this exchange seem simple enough then: an argument is constructed; it is put forward as valid; then it is tested to 'destruction' in the sense that flaws of logic, gaps in evidence, and inaccurate assertions are identified. If no flaws are found, the argument has survived its tests and the system in question will be provisionally accepted for use.

However, most safety experts will recognise this description as extremely idealised. It does not recognise phenomena such as: ambiguities in the interpretation of evidence; professional disagreements about what constitutes 'best practice'; collectively recognised limitations (of knowledge, of technology, of the intractability of certain problems). Neither does it recognise the political, legal, or moral issues involved in safety acceptance; nor changes in public opinion about risk acceptability; nor indeed the question of how (or whether) public opinion is to be elicited in the first place. When so many different forces can be brought to bear on a particular risk decision it seems unlikely that an argument solely based on rational calculation will be sufficient to determine the decision, especially if, as deconstruction implies, the 'meaning effects' of the calculation (as opposed to its more tractable mathematical meaning) will be at the very least problematic.

The challenge that apparently 'objective' models of risk decision-making are really post-hoc rationalisations of asserted biases is not unprecedented in safety engineering. For example, Adams (1995, Chapter 6) argues that since there is no objectively meaningful notion of 'value' upon which to found the monetisation of risk, the results of Cost-Benefit Analysis are determined by the biases of whoever is conducting the analysis. In his words: "cost-benefit analysis is almost always used not to make decisions, but to justify decisions that have already been made".

Similar arguments have been made concerning the use of reliability theory in safety assessment. On this subject, Leveson (1995, p. 168) quotes DT Lowe: "Risk analysis of the type considered here is to safety what the merry-go-round is to transport. We can spend a lot of time and money on it, only to go round and round in circles without really getting anywhere."

Furthermore, problems of constructivism in interpretation often emerge in the context of discussions about Probabilistic Risk Assessment. Crawford (2001) reviews objections to PRA ranging from the subtle to the unsubtle (quoting physicist Richard Feynman as saying "If a guy tells me the probability of failure is 1 in 10^5, I know he's full of crap."). A rejoinder to Crawford's paper provides an amusing example of how different representations of a problem can lead to mutual incomprehension, circular self-justification, and unconscious self-contradiction. Vesely and Fragola (2002) object to Crawford's questioning of the meaningfulness of statistical testing with the words: "The author's conclusions may be interesting conjectures, but they have no statistical basis." This misses Crawford's point entirely, especially as Vesely and Fragola agree that: "*No* PRA should be taken at face value" (our italics).

The phenomenon of circular justification also troubles discussions about risk 'tolerability' (Health & Safety Commission 1998). Our oft-discussed principle is that risks must be reduced to a level that is "as low as reasonably practicable" (ALARP). However, ALARP is based on the assumption that for a particular type of safety system a "tolerable region" actually exists; this guarantees that at least some level of effort will enable it to be attained, but this presupposition is never questioned. The notion of 'tolerability' implies that what is tolerated is undesirable, but in some sense unavoidable; but if risk taking were entirely unavoidable it would not require 'justification'.

We can observe analogous difficulties in discussions about the use of Commercial Off-The-Shelf software in critical applications, particularly where the software in question was not originally developed for critical use. A text that presupposes the possibility of the safety-critical use of COTS and uses this presupposition as the starting point for its arguments will find it impossible to account for phenomena that indicate why COTS use was previously discounted in safety critical systems. For example, one text we have analysed proposes a method for justifying the use of "Software of Unknown Pedigree" (SOUP) in safety-related systems. It alludes briefly to the possibility that 'Easter Eggs' might be buried in COTS software, and comments that "such problems are harder to deal with than in bespoke software …"; but the text seems unwilling to go much further. An alert reader might wonder how a software engineering manager should "deal with" Easter Egg software. Obviously, identification of the offending source code is the first pre-requisite. Subsequently, removal will be desirable. However, this could prove tricky if there are non-functional dependencies between the Easter Egg software and the useful software. It may be that (e.g. due to memory mapping sensitivities) the Easter Egg software cannot be removed; but it could perhaps be rendered harmless by a suitable "wrapper" function.

What of the development process that allows Easter Egg software to get through compilation and build into a shipped delivery? Evidently, the programmers responsible need to be identified and sanctioned. The development process needs to be re-examined and modified to make sure the childish trick will be detected if it is played again. Hence, there is no way to "deal with" a development process that does not permit the identification of individual programmers and is not amenable to the necessary preventative measures. The text avoids detailed consideration of the Easter Egg issue, but when the issue *is* followed up, it quickly leads to general principles that were well-accepted *before* the critical use of SOUP was ever proposed: e.g. that the software source code should be available for inspection; that object code and memory maps should be available just in case; that the software should be open to reconfiguration and recompilation; and that a software development process needs to be traceable and self improving. These principles indicate why, until quite recently, the reuse of SOUP in critical systems was not countenanced. A text that does countenance it can have its 'starting point' 'deconstructed' via the construction of some 'earlier' starting point.

Note that no text can evade this manoeuvre. The text in question gives various hints that the authors were fully aware of the objections to SOUP discussed above at the time of writing; for example, they have included warnings about Easter Egg software in their appendices. However, given the logic of their text and its starting point, they could not make them explicit in the main body. Deconstruction predicts that the author is always to some extent a prisoner of the requirements of their text; the arguments in a text (and this paper can be no exception) are unavoidably shaped by the starting points that have been chosen or specified in advance.

5   Conclusion

A deconstructive reading will be very alert to a text's unwillingness to follow up the implications of avowedly peripheral issues that it does not avoid mention of. There is no reason why safety arguments should not be confronted with the same challenge, particularly as the risks of incompleteness, of 'confirmation bias', and the necessity for creative 'safety imagination' when identifying hazards are already recognised by safety engineers. At the same time there is increasing recognition of the irreducible subjectivity involved in safety judgements and concern over whether 'expert opinion' is enough to cover gaps in meaningful data (Redmill 2002a), (Redmill 2002b), (Adams 1995). Indeed, the controversy that opened up during the drafting of the Royal Society Study Group report on *Risk Analysis, Perception, and Management* (Royal Society 1992) has not receded. Those who think risk management can be given a rational and scientific basis, and those who doubt that it can, seem unable to find arguments that can convince those predisposed to the other view (Hood & Jones 1999).

The deconstructive perspective would view this debate as an inescapable 'social text'. Deconstructive thought predicts that safety engineers will be unable to define any objectively meaningful unit of risk - since meaning is a relational property, there is no 'objectively meaningful' term in any case. Neither will there be any unarguable criteria for determining risk 'tolerability'. Yet we cannot abandon this impossible search without relinquishing our discourse to a self-destructive and indeed contradictory relativism. Rational safety argumentation *requires* basic definitions, but by that very fact remains perpetually vulnerable to a charge of presupposing a conceptual foundation that - if we recognise the challenge of deconstruction - it cannot possibly have. Foundational concepts - for example, the usually unquestioned likelihood-severity distinction - must be artificially constructed; we cannot base them on the 'discovery' of some truth that lies beyond all problems of interpretation.

Deconstruction can be viewed provisionally as a philosophical framework for 'reading between the lines'. It stimulates the readers' imagination by encouraging a close questioning of basic terminology and its effects upon conceptual thinking in what are, after all, rather dry technical documents. Most of the flaws sought out in deconstructive readings are not qualitatively different from those targeted in critical reading generally: e.g. ambiguities, logical fallacies, and gaps in reasoning. Thus deconstructive reading could be used by assessors to analyse safety arguments and used by systems developers in order to test and improve their safety arguments before assessment. However, our examples will have alerted the reader to the fact that deconstructive reading can undermine even the most logically 'watertight' argument by pointing out the limits of the language used to express it. A major question for us is whether such fundamental questioning leads to intellectual paralysis (in what is after all a decision-making process) or whether the insights gained will be intellectually stimulating. Arguably, the point of intellectual paralysis marks where safety 'decision-making' really begins (afresh?).

Acknowledgments: This work was supported under EPSRC research project GR/R65527/01. The authors would like to thank Bob Malcolm for introducing us to the work of Barry Turner on deconstruction and safety engineering.

6 References

Abrams MH (1999). *A Glossary Of Literary Terms*, Seventh Edition, Harcourt Brace College Publishers, ISBN 0-15-505452-X.

Adams J (1995). *Risk*. Routledge, ISBN 1-85728-068-7.

Barthes R (1994). *The Semiotic Challenge*. Translated from the French edition *L'aventure Sémiologique* (Editions de Seuil, 1985) by Richard Howard, University of California Press, ISBN 0-520-08784-4.

Bennington G & Derrida J (1993). *Jacques Derrida*, University of Chicago Press, ISBN 0-226-04262-6.

Cambridge (1999). *The Cambridge Dictionary of Philosophy*, Second Edition, Cambridge University Press, ISBN 0-521-63722-8.

Clark T (2000). *Deconstruction and Technology*. In: Royle 2000, pp. 238-257.

Cobley P (2001). *The Routledge Companion to Semiotics*. Edited by Paul Cobley, Routledge Taylor & Francis Group, ISBN 0-415-243149.

Crawford J (2001). *What's Wrong with the Numbers? A Questioning Look at Probabilistic Risk Assessment*. Journal of System Safety, Vol. 37, No. 3, Third Quarter 2001.

Culler J (1997). *Literary Theory: A Very Short Introduction*. Oxford University Press, ISBN: 0-19-285383.

Culler J (1998). *On Deconstruction: Theory and Criticism After Structuralism*. Routledge, ISBN 0-415-04555-X.

Culler J (2001). *The Pursuit of Signs*. Routledge Classics, Routledge, ISBN 0-4152-5382-9.

Daily Telegraph (2000). *So Just How Safe Is It To Travel by Plane?* Article by Matt Ridley, Wednesday, July 26th 2000. Also available at: http://www.smh.com.au/news/specials/intl/concorde/conair21.html

Derrida J (1967). *Of Grammatology*. Translated by Gayatri Chakravorty Spivak, The John Hopkins University Press, Corrected Edition 1997, ISBN 0-8018-5830-5.

Derrida J (1977). *Limited Inc*. Edited by Gerald Graff, Northwestern University Press, Evanston IL, ISBN 0-8101-0788-0.

Derrida J (1981). *Dissemination*. Translated from the French edition *La Dissémination* (Editions de Seuil, 1972) by Barbara Johnson, Athlone Contemporary European Thinkers, ISBN 0-485-12093-3.

Eagleton T (1996). *Literary Theory: An Introduction*, Second Edition, Blackwell Publishers, ISBN 0-631-20188-2.

Health & Safety Commission (1998). *The Use of Computers In Safety-critical Applications: Final Report of the Study Group on the Safety of Operational Computer Systems*. HSE Books, ISBN 0-7176-1620.

Hood C & Jones DKC (1999). *Accident and Design: Contemporary Debates in Risk Management*, Routledge, ISBN 1-85728-598-0.

Leveson NG (1995). *Safeware: System Safety and Computers*. Addison Wesley, ISBN0-201-11972-2.

Redmill F (2002a). *Risk Analysis – A Subjective Process*. In: *The Engineering Management Journal*, IEE Publications, April 2002, pp. 91 – 96.

Redmill F (2002b). *Exploring Subjectivity in Hazard Analysis*. In: *The Engineering Management Journal*, IEE Publications, June 2002, pp. 139 – 144.

Royal Society (1992). *Risk Analysis, Perception, & Management*, Report of a Royal Society Study Group, The Royal Society, London ISBN 0-85403-467-6.

Royle N (2000). *Deconstructions: A User's Guide*. Edited by Nicholas Royle, Palgrave, ISBN 0-333-71761-9.

Turner BA (1994). *Software and Contingency: The Text and Vocabulary of System Failure?* Journal of Contingencies and Crisis Management, Vol. 2 No. 1, March 1994.

van der Meulen M (2000). *Definitions for Hardware and Software Safety Engineers*. Springer, ISBN 1-85233-175-5.

Vesely W & Fragola J (2002). Untitled article in the *From Our Readers* section. Journal of System Safety, Vol. 38 No. 1, First Quarter 2002, pp 5 – 6.

Ward I (1998). *An Introduction to Critical Legal Theory*. Cavendish Publishing Limited, ISBN 1-85941-348-X.

Wheeler SC (2000). *Deconstruction As Analytic Philosophy*. Stanford University Press Cultural Memory in the Present Series, ISBN 0-8047-3753-3.

THE SAFETY CASE – 1

Developing a Safety Case for Autonomous Vehicle Operation on an Airport

John Spriggs

Safety Assurance Consultant, Roke Manor Research Limited,
Romsey, SO51 0ZN U.K.

Abstract

This paper discusses the development of a safety case for the operation of an autonomous vehicle near to an airport's runways and taxiways. It describes an approach to constraining such vehicles to operate only in allowed areas, and highlights some of the problems that may be encountered in constructing the safety argument.

Introduction

In my paper given at the Safety-critical Systems Symposium 2002, [SSS 2002] I discussed some hazards that may be encountered on airport runways. In particular, I reported on some research concerning the automatic detection of debris and foreign objects on airport runways, and suggested that an autonomous vehicle could be used as the sensor-carrying platform. You may recall that I said that we could build such a vehicle now, using today's technology, but I was doubtful that we could persuade the Safety Regulator that a vehicle guided by machine vision would keep to its designated areas, and not wander onto runways or taxiways without the appropriate clearances. It would need to be fenced in somehow.

The autonomous vehicle cannot be fenced in literally, as the only things that can currently be installed above ground near runways are aids to aircraft navigation. Rail guidance would encounter similar problems to fences, so buried wire guidance to complement the machine vision had been proposed as a solution.

Subsequent private venture work has identified a potential solution that does not involve the disruption inherent in digging a trench and installing wires. In principle, it uses existing safety-critical radio navigation infrastructure. In practice, that infrastructure is currently installed at very few airports; it can, however, be expected to be widely deployed in future.

The autonomous vehicle does not have to be a debris monitor, it could be a bird scarer, or a grass cutter, but whatever it is, we will need a safety case for its operation on the airport. This paper uses the vehicle as an example in discussing how we could construct a safety case. It highlights the fact that re-use of a safety-critical item in a new system does not mean that preparing the associated part of the safety argument will be trivial.

What is a Safety Case?

A safety case in this context is defined as:

"A document which clearly and comprehensively presents sufficient arguments, evidence and assumptions that system hazards have been identified and controlled for both engineering and operational areas to demonstrate that a facility, facilities or organisation is/are adequately safe in air traffic service respects." [CAA 1998]

"An argument is a connected series of statements intended to establish a definite proposition." [Python 1989]

The statements in a safety argument are often referred to as claims. For example a claim may be, "X is safe", for which the terms "X" and "safe" will have been previously defined. Safety is not absolute, so the definition of "safe" will employ words like "tolerable", which must also be defined for a particular application by stating a target level of safety or by stating the degree of acceptability of defined classes of risk.

Evidence is information that is presented to establish the point in question; it supports the claim. If a claim asserts A is true, evidence must be available to demonstrate its truth. Evidence can be brought to support a claim directly, e.g. a safety plan could be presented to demonstrate the claim that all safety activities were planned in advance; but this is not sufficient - backing evidence is also required, e.g. configuration control records and audit reports demonstrating that this is the plan that was implemented.

An assumption is a statement that is believed to be, or taken to be, true for the purposes of the argument. For example, "It is assumed that the existing equipment complies with the pertinent Minimum Aviation System Performance Standards". Evidence should, where possible, be brought to indicate the validity of the assumptions.

The purpose of a safety case is to provide an assurance that any risks that may be introduced by a change to a system or facility have been minimised, as far as is reasonably practicable, before the change is introduced into operational service. The scope of the change may range from the bringing on-line of completely new systems to a minor modification of operating procedures.

Incremental development is very important to fulfilment of this purpose. It is easier to provide assurance to someone who has observed the development of the arguments, than to someone who is suddenly presented with a hundred pages to read. The initial version of the safety case should just present the proposed structure, i.e. the top-level argument, for approval. Subsequent versions would develop the argument further and populate the structure with references out to the supporting evidence. At each stage of development of the item for which the safety case is in preparation, the evidence with which to demonstrate accomplishments should be available and referenced from the safety case. Furthermore, the plans should be in place to collect evidence of accomplishment in the succeeding stages.

How is a Safety Case Presented?

In general a safety case can be a very large document, so we need strategies to break it up into manageable chunks. The first thing to note is that the evidence element of a safety case is often, but not always, material that is also used for other purposes. For example acceptance test results will be used to show that all requirements are fulfilled, not just the safety ones; production test results can be used in the optimisation of manufacturing processes, not just to demonstrate repeatability.

The evidence must be kept elsewhere; not in the safety case document, but referenced therefrom. The whole must be kept under configuration control with all the rest of the system documentation, so that, for example, if a component is no longer available, its replacement is assessed for impact on the safety case before being authorised for use.

Even when keeping the evidence separate you can end up with a large document. You can build a good case like a Victorian novel and convince yourself of its validity because of the process you went through in constructing it. The purpose of a safety case, however, is to convince others, so proof by construction does not work here, unless you also went though an iterative approval process. The safety case is not intended to be a static object, however. It has to develop as the system it describes develops. A safety case with monolithic structure can be very expensive to maintain.

One method of partition that can be used is often expressed by the composite claim "it is safe now and it will continue to be safe", i.e. provide a safety case for entry into service, and one to show how the safety features are preserved in operation. Some authorities explicitly require separate safety case volumes like this.

For our vehicle, we could split the case up into two volumes; the first addressing the development of the safety requirements and their fulfilment in the design, the second addressing the hazards arising from the use of the vehicle and how the mitigations are preserved throughout its life. The first volume is clearly the responsibility of the equipment supplier, whilst the second volume will become the responsibility of the Operators at handover, and will require their input during development.

These volumes address functional safety but, as the designer, there are other aspects that I must consider, for example, the hazards to maintenance personnel, their colleagues and their visitors arising from the vehicle. These are "product safety" considerations, and deserve their own volume. Even though the Regulator is unlikely to ask me about them, the Operators will need the assurance and will take over responsibility for this volume also, unless maintenance is contracted out.

Furthermore, I must ensure that the vehicle is safe to build and test. Production test personnel, their colleagues and visitors, may be open to hazards not encountered in normal use or at first line maintenance because, for example, they may be working on a subassembly rather than the whole thing, and they may be using high power simulated signal sources, etc. In practice, I would slip this topic into the safety case as an extreme form of maintenance, but it has to be explicit, so my top-level set of claims, which define the partition into separate volumes, becomes:

- The vehicle is safe to enter service

- The vehicle remains safe in operation

- The vehicle is safe to manufacture and maintain

These claims can be further decomposed using arguments, to support each claim, that in turn depend on sub-claims. We proceed in an iterative manner to populate each volume of the safety case. We need to address many topics in this decomposition, for example, design processes; quality assurance; test strategies; staff competence; the derivation, validation and verification of safety requirements; configuration control; analyses of unwanted interactions, interference and misuse; the need for special operating procedures...

Right at the bottom of the tree we would have claims that cannot usefully be further broken down and just point to supporting evidence, like:

- The Product Safety Review was completed with no actions arising.

- CE Marking Declarations of Conformity were obtained for Machinery, EMC, RTTE and Low Voltage Directives.

It is not enough to have a sensible decomposition, we need also to illustrate the structure so that the reader can understand it and navigate through it. A cunningly crafted contents list could structure the claims; but this only really works for small systems. A picture is much better at conveying a tree structure than a list; but that is not enough either.

A Safety Case is Not Just What You Have Done

The Regulatory Authority will want to know why you have chosen this structure; in particular, they want to know why you think that doing it this way is sufficient to fulfil the objectives that they have set. Decisions need to be justified in the safety case; it is not enough to claim:

- The software was developed to Class C of Standard X

You also need to state why that particular standard is thought appropriate and, in the light of that, why you have chosen to address the objectives of Class C, rather than Class B, say. I once asked why a particular standard had been chosen for a project and was told something like "our avionics colleagues use this for aircraft control systems, so it must be good enough for our train application". If you have to justify your choice of standards explicitly, i.e. produce an argument to illustrate how the standard is appropriate, you may think about it more deeply than this; you may even come up with something more cost-effective. Similarly, the use of particular tools needs to be justified and evidence brought to show that they have been verified.

"But the standards and tools were chosen at bid time, whereas the safety case is the last deliverable and has to reflect what has been done; it is a fait accompli!" If you are in the unenviable situation of having to bid for a project to develop a safety–related system, with no intermediate feasibility studies, risk reduction exercises, demonstrators, etc., I suggest that you do not produce the safety plan that the Customer has requested in the tender documents. Instead, produce Issue A of the safety case; it serves the same purpose, but includes justification of the plans.

This first iteration of the safety case is all intentions rather than claims – "it will be safe because..." This gives you the opportunity to build the structure of the claims and the arguments, at least at a high level. Most importantly, it allows you the opportunity to provide design constraints, with justification, to the systems engineers who are developing the architecture and draft design specifications.

I am sure you have met project managers who call for safety arguments to demonstrate that what has been done is adequate for the application. They think it your fault if it turns out to be inadequate; and, of course, it is your fault if you were present at the planning stage and did not consider the impact of the decisions on the safety case, and vice versa.

What this digression illustrates is that we need to state justifications, i.e. arguments of why a claim is thought appropriate, as well as the arguments that link the claims to sub-claims. We also need to represent the relationship between all the various elements of a safety case in a clear manner.

I have looked at using the Ward and Mellor Entity Relationship Diagram [Ward 1987] as a means of capturing the structure of a safety case – the aim was to use existing software tools, and also to help build the safety case in a data base format. This work has been overtaken by events; Ward and Mellor have fallen out of fashion in the software world and, more significantly, special purpose tools and notations have now become available, for example the Adelard Safety Case Editor and the Goal Structuring Notation from York University.

Goal Structuring Notation allows construction of complex safety cases whilst making explicit the logical relationships between the claims (Goals) sub-claims (Sub-goals), strategies, rationales, contexts and evidence. This is a very powerful notation because it captures context information and the assumptions made in developing the argument. The capture of this information opens the way for future re-use of parts of the safety argument, it is key to establishing whether an existing argument may be re-used in a new application.

In the simple example of Figure 1 overleaf, which uses a sub-set of the notation, goals and sub-goals are represented by rectangles; the context information is in rectangles with rounded corners; a rhombus is used for strategy; and sources or types of evidence are circles. Of course, a real safety case would further decompose the sub-goals before appealing to the evidence. There are a number of choices for this decomposition, for example the split could be on the basis of particular rôles and functions; or it could be on the basis of personnel, procedures and platforms; or it could even be to maximise re-use of existing safety arguments.

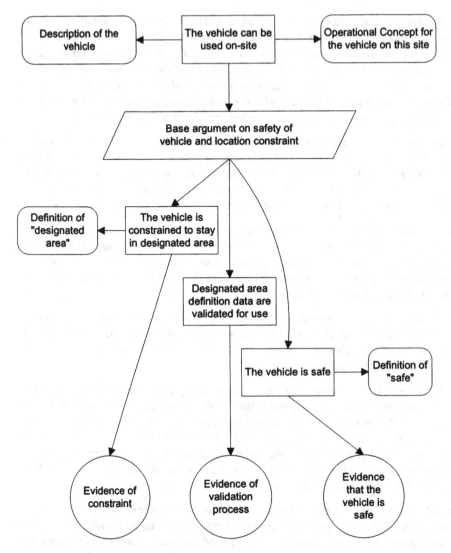

Figure 1 ~ Simplistic Goal Structuring Sub-network for the Safety Case

We will opt for a functional decomposition of the claim "The vehicle is constrained to stay in designated area", giving sub-claims:

- The vehicle has a definition of the designated area
- The vehicle has a predefined operational strategy
- The vehicle has access to a high-integrity navigation service
- The vehicle uses received navigation data to direct and bound its operations.

The decomposition of the first point would then go into implementation detail, discussing the definition of the boundary of the area in question plus a guard band defined by protection limits. If the vehicle guidance detects that it is approaching a protection limit, it has to react so that factors like momentum and uncertainty in position do not act to push it over the boundary inadvertently.

Reusing an Existing Safety-critical Item

Recall that the constraint keeping the vehicle off the runway was to make use of an existing infrastructure item. What we propose to use are the radio navigation signals from a "blind landing" system, i.e. one that is used in the landing of aircraft when there is insufficient visibility for the pilot to do it "manually". Some would say that there is no issue here; we have a system certificated to land aircraft with hundreds of passengers automatically, so it must be safe enough to mow the grass!

Well, no, it does not work like that.

You could expect that one of the main thrusts of the landing system safety case would be concerned with the continued ability to keep the aircraft tracks coincident with the centre line of the runway (plus or minus a little bit). This is just what we want the grass cutter not to do. I may be able to appeal to the same body of evidence about the infrastructure, but I certainly cannot use the same argument in both cases.

This is the main lesson I want you to take from this paper. Just because you are re-using part of an existing safety-critical system in your new system, it does not mean that the pertinent part of the safety argument will be trivial. We have an advantage in this particular instance, in that we are re-using an established, well-documented, interface – but that does not stop the analysis from being horrendously complex. I will give an overview to illustrate that point.

In fact, the landing systems are so well defined that any safety case is more likely to argue from the viewpoint of compliance with standards, than it is to start from a requirement to stay on the centre line, or the avoidance of obstacles on approach.

Standardisation of Interfaces for Interoperability

The first thing to note is why the interface that we wish to exploit is well documented. It arises from the nature and structure of the industry and its Regulatory environment. A hierarchy of standards, specifications and characteristics has to be in place before the detailed design, development and entry into service of equipment required to implement a new aeronautical service.

By its very nature, aviation operates across national boundaries; aircraft are required to operate globally in a safe and expeditious manner. Furthermore, different companies produce the equipment for the ground-based and airborne segments; and they do this largely independently, based on the requirements of the standards. International agreements are made apportioning safety performance requirements to the subsystems. In particular, Standards and Recommended Practices (SARPS) are developed under the auspices of the International Civil Aviation Organization (ICAO), a United Nations agency.

SARPS normally address essential air-ground interface parameters, e.g. signal strengths in coverage areas, power output, frequency, tolerances, modulations, signal protocols, integrity and continuity of service. The treatment of the ground segment in distributed air-ground systems is usually more comprehensive and detailed than that for the airborne segment. This emphasis is intended to encourage a worldwide uniformity in the provision of services. It also assists countries with less well-developed infrastructure to procure their ground systems with a high degree of confidence as to their appropriateness to provide an adequately safe and expeditious service.

SARPS are developed through specially constituted specialist panels and working groups. Once representative drafts are available from the iterative development process within the working groups, they are submitted to the panel and, if acceptable, onward for worldwide discussion and agreement at ICAO Divisional meetings. They are published, following a written consultative procedure, in annexes to the Convention on International Civil Aviation. This requires the assent of a majority of member states.

Blind landing systems are considered a sub-set of Telecommunications Systems, and are standardised in Volume 1 of Annex 10 to the Convention [ICAO 1996].

For a typical landing system, there are defined three main parameters relating to safety of operation:

Continuity of Service It has to be operating throughout the period in which you need to use it, i.e. approach and landing

Integrity It either has to give you correct information, or tell you that it is not correct

Accuracy It has to give you the information required for a soft landing at the right place on the runway

Continuity of Service is, in effect, reliability of the navigation service and it does not really affect us in this application. Our argument will be based on the principle that, if the service stops, the vehicle will be commanded to stop and so it remains in its permitted operating area. Continuity of Service is, of course, very important for the landing application, as the aircraft is not in a position to stop. Procedures define a decision height above which they can elect to abort the landing and go around to begin another approach; below that height, they have to land.

Integrity is important to our vehicle; if it is in one place, but the navigation service suggests that it is somewhere else, it could encroach onto forbidden areas. It would not be a sudden change, as that would be readily detectable, but a gradual drift. Accuracy is also important; if the navigation service tells us where we are plus or minus a metre, then we can accept that; if it is plus or minus fifteen metres, we would need such a large guard band around our operating area that, for example, much of the grass would not get cut.

It is not enough to take numbers from the standard and use them in our arguments. To do that would be to make an implicit assumption that the numbers are appropriate for use in this way. We need really to understand what they mean, and justify their application. I originally assumed, for example, that the Integrity requirement would be the same as that I had encountered in my work on Microwave Landing Systems some years ago, but it is not. Integrity is defined for landing systems as:

"That quality which relates to the trust which can be placed in the correctness of the information supplied by the facility".

The original Instrument Landing System and the more recent Microwave Landing System are defined to have radiation monitors and functions that are arranged to stop transmission if pre-defined thresholds are exceeded.

Integrity is expressed as one minus the probability of transmission of undetected erroneous radiation. That probability is that of concurrent failures in transmitter and monitor resulting in undetected erroneous radiation. It can be calculated from a knowledge of the Mean Times Between Failure of the items and of the proportion of those failures that can cause the emission of spurious guidance.

The original landing systems used only hardware components in the transmission and monitor paths, and so the techniques of traditional Reliability Analysis can be used for these calculations. Some more recent systems have an element of software in the monitoring functions; as there are no recognised techniques for the development of software to the required integrity levels, additional hardware functions have to be incorporated to apply checks on the results provided by the software components.

The system that we wish to use is the Ground Based Augmentation System, which is also known as the Local Area Augmentation System in some parts of the World. It is different from the original landing systems in that the ground segment does not provide the navigation signals per se; there is an additional, space-based, segment that does that. The ground segment acts as a monitor; but in this case, it does more than just compare against thresholds. It provides corrections and other data needed to produce a dependable navigation solution. These additional data include parameters that are to be used by the aircraft to check the integrity of its navigation solution.

This is a real system; installations are planned or already in place at, for example Frankfurt Main, Chicago O'Hare and Memphis International airports. Many others will follow, especially as the original instrument landing systems approach the end of their design life.

To consider the meaning of integrity of the service provided by the Ground Based Augmentation System, we need to know how it works and what can cause erroneous results.

Satellite Navigation

The concept of Satellite Navigation using, for example, GPS[1] or GLONASS[2], is well known; what is probably less well known is that such systems are inadequate for use by commercial air traffic. Although the accuracy obtainable may be adequate for en-route navigation, the Integrity and Continuity of Service performance of these systems leaves a bit to be desired. There are, for example, a number of mechanisms for undetected corruption of the navigation solution. These factors would be problems for "blind landing" too, but the dominant problem there is the fact that the systems are just not sufficiently accurate, especially in the vertical axis. However, before mitigations of these problems can be discussed, we need to know how the satellite navigation systems work.

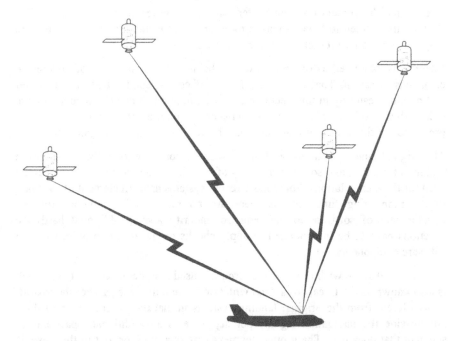

Figure 2 ~ Basic Configuration Required for Satellite Navigation of Aircraft

GPS and GLONASS allow positions on, or near, the Earth to be estimated by using multilateration to a number of orbiting satellites. The distance to each satellite in view is measured by timing how long it takes for a signal transmitted by the satellite to arrive at the receiver. The distance is then the product of the measured period and the speed of light, which is defined to be $2{\cdot}99792458 \times 10^8$ m.s^{-1}.

---

[1] The Global Positioning System

[2] The Global Navigation Satellite System

The theoretical geometry is relatively straightforward. If the distance from one satellite is known, then all that can be deduced is that the receiver lies somewhere on a sphere centred on the satellite with radius approximately twenty thousand kilometres. Two transmitters bring it down to a circle, and three to a pair of points. Four transmitter distances give an unambiguous result; although three could be sufficient, as one of the pair of points would be inappropriate for an aircraft, e.g. outside the atmosphere, or underground.

In practice, however, four distance measurements do not give an unambiguous result. The distance is being measured by timing, but different clocks are being used at either end of the links; consequently, the spheres do not intersect at a point. The receiver clock will not in general have the accuracy and stability of the satellite system, which in GPS depends on an ensemble of two caesium and two rubidium atomic clocks. However, a single receiver makes all the measurements, and so, for first order effects, there is a single value of "clock error" that needs to be added to each of the measurements to achieve intersection.

Once that correction factor has been determined, the receiver can, in principle, apply it to all measurements from then on. In effect, the receiver clock is now synchronised to satellite time, which has a known relationship to Co-ordinated Universal Time (UTC). In practice, the correction process is usually repeated for each set of range measurements taken, to ensure synchronisation is maintained with varying clock error. A receiver does not use the clock correction to actually correct its clock.

The uncorrected range measurements are called "pseudo-ranges" to indicate that they are measurements that contain errors. In order to identify reference points on the signal for timing measurements, pseudo-random code sequences with excellent autocorrelation properties are modulated onto the carrier. Each GPS satellite has its own code and so they can transmit on the same frequency without interfering with each other. GLONASS is the other way around; all the satellites use the same code, but are each allocated a different transmission frequency.

The navigation solution depends on knowing where the transmitters are as well as "when they are". Data with which to calculate the position of the satellite when the signal was transmitted are contained in messages superimposed on the pseudo-random codes. These messages also contain the time of transmission, plus data for error reduction and availability data for the satellites. The satellite position problem is simplified by the high altitude. They are well clear of the atmosphere, and so have readily predictable orbits. The ephemeris giving satellite position, with a lower precision almanac of all the satellites, is also transmitted.

In summary, the receiver derives its own position from knowledge of the position of a number of satellites from which it receives signals. Furthermore, it is able to derive a correction factor for its clock. What is usually done in practice is to derive correction factors for all dimensions; the receiver maintains an estimate of where, and when, it is, which is updated as each set of measurements is made. A Kalman Filter implementation is almost exclusively used to maintain the estimates.

As the system depends upon taking measurements, there are errors to be taken into account. There are, of course, errors in the receivers themselves, e.g. imprecise phase measurement, but these are reduced by design in high-quality equipment to affect the navigation solution by less than two hundred millimetres. There are much larger errors arising from our assumption about the transmission path.

Sources of Error in the Signal Path

One source of error should be apparent from the previous description; the speed of light is defined as a particular number. That number is derived from the Système International definition of a metre:

"The length of the path travelled by light in vacuum during a time interval of 1/299792458 of a second".

The definition acknowledges the axiom of special relativity that says that the speed of light is a constant. That is the speed of light in a vacuum; these signals travel through the atmosphere, and so are slower. The ratio by which they are slowed is the refractive index. One would expect the refractive index to increase along the signal path, as one end is in effective vacuum, and the other is near the base of the atmosphere.

Unfortunately, it is not that simple; the upper part of the atmosphere, the Ionosphere, has an effect on signal propagation, due to the presence of charged particles, that varies over time. The Ionosphere is usually reasonably well behaved and stable in the temperate zones; but near the equator or magnetic poles, it can fluctuate considerably. It has a greater effect during the day than at night; there is also a long-term periodic component, apparently correlated with the sunspot cycle.

Some GPS receivers use a model of the Ionosphere, with parameter data provided by the satellites, to derive a correction. Residual errors in the final result tend to be in the region of two to five metres despite this correction.

Military users of GPS have access to another signal transmitted from each satellite on a different frequency. As the ionospheric delay effect is proportional to the frequency of the signal, tracking both signals allows the delay to be calculated and hence compensated for. There are second order effects; residual errors in receivers using the dual frequency technique are of the order of one or two metres.

Lower down in the atmosphere, the Troposphere has different delay characteristics. These are due primarily to the water vapour content, but there are also temperature and pressure effects on both code and carrier. These are not frequency dependent; but it is feasible to compensate for the effect by measuring these parameters and applying them to a tropospheric model to derive a delay estimate. In practice the tropospheric models have been found to be better than those of the Ionosphere, but there may still be errors in the final solution of the order of a metre.

There are error-producing effects in the system that are more esoteric in origin.

Relativistic Effects

One of the consequences arising from the constancy of the speed of light is that a moving clock appears to run slow with respect to a similar clock that is at rest. The GPS satellites travel in their orbits at such a speed that their clocks appear to run slow by seven microseconds per day, relative to a similar clock on the Earth's surface.

General Relativity tells us that a clock at a weaker gravitational potential appears to run fast in comparison to one at a stronger potential. This effect makes the satellite clocks appear to be running fast by forty-five microseconds per day. The net effect is thus that the satellite clock is fast, by thirty-eight microseconds per day, with respect to a clock on the ground. This systematic error is compensated for in GPS satellites by a rate offset introduced into the satellite clock before launch. The master oscillator runs at 10·22999999543 MHz, which appears to be the required 10·23 MHz when in orbit.

The orbits are not exactly circular, so the altitude varies over one revolution, as does the speed. The eccentricity of the orbits gives an effect in addition to that compensated for by the fixed rate offset. Data transmitted from the satellite includes additional coefficients to enable the receiver to model, and compensate for, the effect.

There is a third relativistic effect to be taken into account. A clock on the Earth is not in fact fixed; it is in a rotating frame. The receiver undergoes a displacement during the signal propagation period. This phenomenon, called the Sagnac Effect, can also be compensated for in the receiver.

Other Sources of Error

Another error contribution is due to the signals being reflected off objects in the environment and then to the receiver. If the indirect path is not a lot different from the direct one, it produces an uncertainty in the arrival time measurement. If the path lengths are very different, the effect may be reduced by signal processing. It is also possible to improve multi-path effects by choice of antenna location and arranging it to reject signals from low angles, from where most of the reflections come. It is possible, but very rare, to encounter ranging errors up to fifteen metres due to multi-path effects; this may happen in a scenario where a static antenna is mounted near large reflecting surfaces. The effects are much less pronounced in moving receivers, and one can assume that multi-path effects are very much less for an air vehicle than for a ground-based receiver, however they can increase as the aircraft approaches the runway.

Unlike GLONASS, GPS has intentional loss of integrity designed into the system; the signals available to non-military users, until recently, have been intentionally degraded. This was done by dithering the satellite time value and the position data. This feature, Selective Availability, was turned off by Presidential Decree in 2000, but there is no guarantee that it will not be turned back on again in future. If it were, military GPS users could obtain the "correct" values; however, there would still be errors due to satellite clock drift and orbit perturbations.

One nanosecond error in the satellite clock corresponds to about three hundred millimetres error in the range measurement. The atomic clocks on the GPS satellites drift and can easily accumulate errors of this magnitude. The US Department of Defense has facilities that monitor the satellite clocks and compare them with a master clock comprising an ensemble of very accurate clocks including, for example, hydrogen masers. Correction factors are then transmitted up to the satellite for inclusion in its transmissions. The receiver, in principle, subtracts the reported error from the transmit-time value to arrive at its navigation solution. Both civilian and military users are affected by this error, which can offset the solution by one or two metres.

Similarly, the satellite orbits are monitored and the data contained in the messages are updated accordingly. So called Ephemeris Errors result when the message does not contain the correct satellite location. It is typical that the radial component of this error is the smallest: the tangential and cross-track errors may be larger by an order of magnitude. Fortunately, these larger components do not affect ranging accuracy to the same degree. Even so, this error source can still contribute a few metres of error to the navigation solution.

All the errors discussed above have direct effects upon the measurement of the satellite ranges. The effect of the range errors on the navigation solution depends on the geometry of the situation. If only the four satellites discussed above are used, and they are clustered in the same part of the sky, the position error can be tens of metres for one metre of range error. Conversely, if many satellites are used, and they are distributed across the sky, then the position errors will be only slightly greater than the range errors.

Differential Mode of Operation

As well as compensating for the various identified errors directly, we can take advantage of the continuity (in the mathematical sense) of the error effects. If two receivers are relatively close, the ranging errors due to propagation conditions and the uncertainties in satellite position and clock will be very similar. If the position of one of the receivers is known precisely, it can calculate what pseudo-ranges it expects for each satellite in view. It can then measure the pseudo-ranges as received and, by subtraction, derive the magnitude of the errors.

Correction values can then be supplied to the second receiver with which to improve its navigation solution. This is more flexible than measuring a position and comparing it with the known position, as it does not constrain the second receiver to use exactly the same set of satellites as the reference to calculate its position. This also allows the second receiver to be mobile within the coverage area of the reference receiver. The error magnitudes change quite rapidly due to satellite motion, and clock drift, so the correction values have to be continuously updated. This is usually achieved via a radio link between the systems; but where position is required as part of some later analysis, the pseudo-ranges and corrections may be recorded, locally to where they are measured, for post-processing "off line".

Multi-path and internal receiver errors are local to each receiver, and so cannot be compensated for in this way. Multi-path dominates for a high quality receiver but, as previously noted, it can be reduced by signal processing and antenna design. For a fixed reference receiver, known sources of multi-path may be mitigated using stealth-engineering techniques to lessen their effects.

Individual pseudo-ranges must be corrected prior to the formation of a navigation solution. The reference receiver therefore needs to track all satellites in view and derive individual pseudo-range corrections for each of them. Similarly, the remote receiver has to be capable of applying the corrections to each set of satellite measurements used to form its navigation solution.

This is in principle how the Ground Based Augmentation System, illustrated overleaf, works but, as I keep on saying, it is not that simple. Variants of this system are defined with two, three, or four carefully surveyed-in reference stations, rather than just the one. These can perform crosschecks on each other, as well as providing corrections and other data to the aircraft via a VHF datalink. In common with the original landing systems, there will be a radiation monitor associated with the datalink transmitter to improve the integrity of the link.

Some implementations, in which continuous view of four or more satellites cannot be guaranteed, also employ "pseudolites", which are ground based signal sources that simulate navigation satellites. The pseudolite provides continuity for the case in which one of only four satellites in view fails during a landing operation. There are also other signal sources that may be used by an installation. The Space Based Augmentation Systems intended for en-route navigation employ geostationary communications satellites in place of the VHF datalink; these satellites also provide navigation transmissions conforming to the GPS standards.

Integrity

Integrity for this navigation service, then, is not just a property of a ground-based transmitter and a ground based monitor, as in the previous landing systems. For this system, it is an emergent property of:

- A navigation satellite constellation, with its own ground-based monitoring and control system;

- The 'signals in space', with all their tribulations;

- The ground-based monitors with their navigation software;

- The ground-based processors implementing the alerting functions; and

- The VHF datalink, with its own radiation monitors

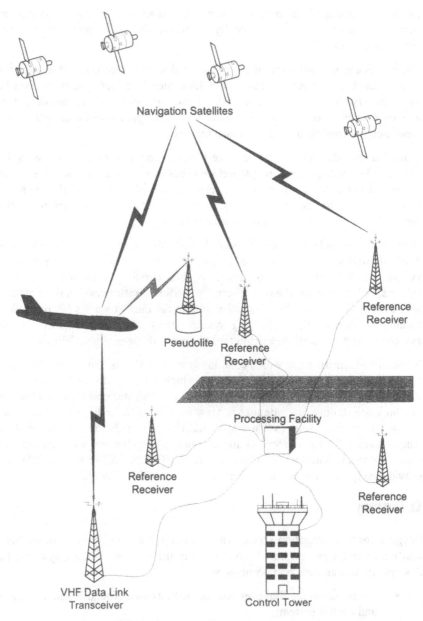

Figure 3 ~ A Ground Based Augmentation System Architecture

The supplier of the ground equipment has a problem. The numbers specified as acceptable for the various integrity-related probabilities are very small; it is not feasible to demonstrate that they have been achieved in the timescales of a normal development programme. They have to adopt other strategies, such as appealing to analysis and simulation. Their safety case has to justify and develop these strategies and provide evidence to validate the simulation results, etc. Furthermore, not only is the integrity calculation more complex than for the original landing systems but also, because of the need to employ software for the Kalman Filters, etc., traditional reliability prediction techniques cannot be directly applied in its verification.

Fortunately, as a user of the service, we can regard this work to have been done. We will state an assumption in our safety case that the evidence seen by the Regulatory Authority, when accepting the landing system for service, was enough to convince them that the safety performance against the pertinent parameters had been adequately demonstrated. Our safety case is thus dependent upon that of another supplier who, in general, will not be willing to show it to us. That is another reason why we cannot re-use parts of the argument in our safety case, and why we need the Regulator to maintain the "web of trust".

For critical infrastructure items, we should not be able to view the safety case even if the supplier had no objections. These security critical documents highlight vulnerabilities in the infrastructure and should remain confidential.

Elaboration of the Integrity Sub-claim

Having seen the Ground Based Augmentation System description, we are now in a position to detail the sub-claim stated as:

- The vehicle has access to a high-integrity navigation service

A candidate structure in Goal Structuring Notation is presented for this sub-claim in Figure 4 overleaf. Although use of acronyms in safety cases should be deprecated, the acronym for Ground Based Augmentation System has been used in Figure 4 to reduce the size of the text boxes.

The new sub-claims assert that the vehicle has the means to receive and use the navigation service and that it is constrained to operate within the local area in which the service is available. The certification of the system will apply over a particular volume but it is, of course, concentrated on aircraft operations. There may be areas on the airport in which we require to operate the vehicle but to which aircraft would, or could, not go. An airport may be perceived as a wide open flat area, but there can be some very large buildings producing the same "urban canyon" effects that cause problems with conventional GPS receivers in cities. Hence the need for signal surveys.

Note that, in Figure 4, the term "designated airport" has been used; this is part of a generic safety case for the vehicle, not for a particular instance. The signal survey evidence will be plans, including methods, for the particular type of survey. We have a software tool for predicting the urban canyon effect, so simple surveys would be used to verify the predictions, rather than having to do a complex series of measurements.

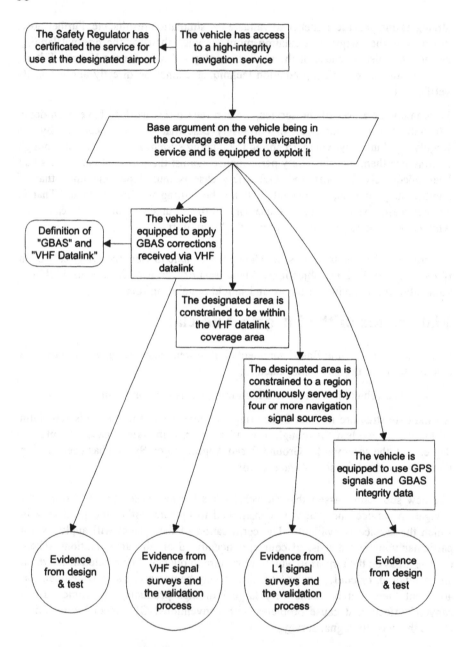

Figure 4 ~ A Breakdown of the High-integrity Sub-claim of Figure 1

One of the sub-claims refers to the integrity data; these data were glossed over in the preceding description. It is not only pseudo-range corrections that are received via the VHF datalink, there are, inter alia, additional statistical parameters that can be used, by the receiver, to judge the quality of the signals prior to use and invalidate the guidance data output if necessary.

Conclusions

A safety case should be built and reviewed incrementally to facilitate approval. It also needs to be presented in such a manner that, in addition to giving confidence to the Regulatory Authority that what has been done is sufficient, it can:

- Give confidence to the organisation producing it that what they have done is necessary;

- Make it straightforward to assess the impact of future proposed changes to the subject system;

- Be easy to maintain; and

- Be used as the basis for future safety arguments in the domain

When budgeting for the safety case development activity, we must not underestimate the amount of work involved in re-using existing items. Those items that are "owned" by the designer, and already have arguments developed and represented in a way that preserves context and assumptions, can be straightforward to assimilate.

However, as illustrated in this paper, incorporation of items that are owned elsewhere, or are to be used in a different context, can involve a lot of effort. We were fortunate in this case that we were tapping in to a very well defined interface. It is also a very public interface in that the International Civil Aviation Organization has published its description. This description has been open to scrutiny by many experts from around the World, and has been made more robust as a result of their work.

Outlining a safety argument in terms of intentions at bid time will help to identify where in-depth analyses will be required, and hence will lead to a more robust estimate of the costs involved in safety case development.

References

[CAA 1998]   Civil Aviation Authority: CAP 670: Air Traffic Services Safety Requirements, CAA London, 1998, plus amendments.

[ICAO 1996]   International Civil Aviation Organization: International Standards and Recommended Practices ~ Aeronautical Telecommunications; Annex 10 to the Convention on International Civil Aviation, Volume 1: Radio Navigation Aids, Second Edition. ICAO Montreal, 1996, plus amendments.

[Python 1989]   Chapman G, Cleese J, Gilliam T, Idle E, Jones T, and Palin M: The Argument Clinic, in Volume 2 of Monty Python's Flying Circus ~ Just the Words, Methuen 1989.

98

[SSS 2002]   Spriggs J: Airport Risk Assessment ~ Examples, Models and Mitigations, in Redmill F, and Anderson T: Components of System Safety, Proceedings of the Tenth Safety-critical Systems Symposium, Springer-Verlag, 2002.

[Ward 1987]   Ward P, and Mellor S: Structured Development for Real-Time Systems, Prentice Hall, 1987.

Managing Complex Safety Cases

T P Kelly

Department of Computer Science

University of York, York, YO10 5DD, UK.

tim.kelly@cs.york.ac.uk

Abstract

Safety case reports are often complex documents presenting complex arguments. To manage the complexity of safety case construction, system safety cases are often decomposed into subsystem safety cases. In this paper we discuss the motivation and problems of partitioning the safety case, both as practiced historically, and as required in new modular, reconfigurable systems such as Integrated Modular Avionics. Recent work on managing safety cases "in-the-large" is presented. In particular, we demonstrate how notions of software and systems architecture design can be read-across to establish the concepts of "safety case architecture" and contract based reasoning for managing inter-safety case dependency. Problems of division of responsibility in safety case development will also be discussed.

1 Introduction

Safety case reports are often complex documents presenting complex arguments. Very rarely is it that safety cases are prepared by individuals. The reality is that the activity of establishing a safety case will be divided amongst a number of individuals, teams and, in some cases, organisations.

To manage the complexity of safety case construction, system safety cases are often decomposed into subsystem safety cases. Many examples of this can be observed in current safety-critical systems: System safety cases incorporate software safety cases (a division advocated by U.K. Defence Standards 00-55 (MoD, 1997) and 00-56 (MoD, 1996)). A safety case concerned with the avionics of a complex military aircraft will be split into separate safety cases for separate systems (such as the navigation system, engine control and flight control). The implied safety case for UK rail operations is made up of separate safety cases for station operations, infrastructure and rolling stock. However, it is well understood that safety is a system property. Extreme care must therefore be taken to ensure that safety case boundaries are drawn correctly, that arguments don't "fall between the gaps", and that formalising boundaries (e.g. through contractual agreements between organisations) doesn't prevent the development of efficient safety solutions.

Emergent safety properties, not dealt with by a "Divide and Conquer" approach to safety case construction, must also be addressed.

In the field of civil engineering if we compare the two cases of attempting to build a skyscraper and attempting to build a tree house we understand that fundamentally different approaches are required. Whilst it may be acceptable to start building the tree house with no clear idea of the overall design and by adding each piece at a time, building the skyscraper requires a clear view of the overall building architecture and guiding principles of construction. In the fields of systems and software engineering the role of *architecture* as a means of managing complexity and achieving emergent qualities such as modifiability is increasingly well understood.

In this paper we demonstrate how many of the principles of systems and software architecture can be carried across to the activity of safety case management in order to help manage complex safety cases.

2   Safety Case Architecture

In the field of software engineering, software architecture has been defined in the following terms (Bass et al., 1998):

> *"The structure or structures of the system, which comprise software components, the externally visible properties of those components, and the relationships among them"*

Safety case architecture can be defined in very similar terms:

> *The high level organisation of the safety case into components of arguments and evidence, the externally visible properties of these components, and the interdependencies that exist between them*

Being clear of the externally visible properties of any safety case module (or 'component') allows us to appreciate its role within the overall safety case structure. The following can be regarded the key 'interface' properties for any safety case module:

1.   Objectives addressed by the module

2.   Evidence presented within the module

3.   Context defined within the module

It is important to note that the definition of safety case architecture provided above gives equal importance to the dependencies between safety case modules (or 'components') as to the components themselves. This thinking must be at the heart of any attempt to decompose the safety case. Safety is not a "sum of parts" property. Dependencies must therefore also be recorded as part of any interface definition, perhaps along the following lines:

4. Arguments requiring support from other modules

5. Reliance on objectives addressed elsewhere

6. Reliance on evidence presented elsewhere

7. Reliance on context defined elsewhere

Safety case modules can be usefully composed if their objectives and arguments complement each other – i.e. one or more of the objectives supported by a module match one or more of the arguments requiring support in the other. For example, the software safety argument is usefully composed with the system safety argument if the software argument supports one or more of objectives set by the system argument. At the same time, an important side-condition is that the collective evidence and assumed context of one module is consistent with that presented in the other. For example, an operational usage context assumed within the software safety argument must be consistent with that put forward within the system level argument.

Where a successful match (composition) can be made of two or more modules, a contract should be recorded of the agreed relationship between the modules. This is a commonplace approach in component based software engineering where contracts are drawn up of the services a software component *requires* of, and *provides* to, its peer components, e.g. as in Meyer's Eiffel contracts (Meyer, 1992).

There are many reasons why it is desirable to break down safety cases into smaller safety case modules for their development and presentation. Firstly, breaking down the safety case into separate components makes concurrent and separate development of the overall safety case (by different teams) possible.

Secondly, it can be useful for isolating change. It is desirable that when changes are required to the system safety argument (either because of challenging evidence, a changing design or a changing regulatory context (Kelly and McDermid, 2001)) we can avoid revisiting the entire safety case. In software component contracts, if a component continues to fulfil its side of the contract with its peer components (regardless of internal component implementation detail or change) the overall system functionality is expected to be maintained. Similarly, contracts between safety case modules allow the overall argument to be sustained whilst the internal details of module arguments (including use of evidence) are changed or entirely substituted for alternative arguments provided that the guarantees of the module contract continue to be upheld.

3 Requirements from Existing Safety Standards

Whilst the approach of 'modularising' the safety case may appear novel, in this section we will describe how the principles are implicit in a number of existing safety standards.

The U.K. Ministry of Defence Standards 00-55 (MoD, 1997) and 00-56 (MoD, 1996) talk of a division of a safety case across the system and software boundary.

Defence Standard 00-56 demands the production of a System Safety Case for all new systems, modifications of existing systems and non-development items. For safety-related software in defence equipment Defence Standard 00-55 requires the production of a *Software* Safety Case:

> "***Part 1 Clause 7.1.1*** *The Software Design Authority shall produce a Software Safety Case as part of the equipment safety case defined in Def Stan 00-56.*"

The role of the software safety case as forming part of the system safety case is clearly recognized. The software safety case must support the objectives of the system safety case. The standard reinforces this point in the following guidance provided in Part 2 Annex E:

> "*E.2.4 Relationship to System Safety Case*
>
> *E.2.4.1 Safety is a system property, and it is impossible to present safety arguments about software in isolation. The Software Safety Case should, therefore, be part of and subordinate to the System Safety Case. Traceability from the Software Safety Case to the System Safety Case should be provided for safety requirements and safety arguments.*"

The principle described above of maintaining an account of the traceability between the system and software safety cases is in agreement with the notion of recording contracts between safety case modules as described in the previous section. The interface of any (sub) safety case must be clearly discernable and the 'contract' of agreement between this safety case and others clearly recorded. Defence Standard 00-55 describes further the form that this contract might take between the system and software case elements:

> "*E.3.6.1 The System Safety Case will also impose 'derived requirements' on the software and ... the Software Safety Case should consist of a number of claims which link back to requirements and constraints imposed by the System Safety Case.*"

Going further than the division of the overall safety case into system and software elements, 00-55 acknowledges the possibility of having to further subdivide the software safety case into separate elements, as shown in the following clause from Part 2 Annex E:

> "*E.2.5.1 For extremely large or complex software systems, it may be necessary to structure the Software Safety Case in line with the architecture of the software.*"

We will describe later in section 6 some of the various 'styles' of structuring and subdivision of the safety case.

Moving away from the particular division of system and software safety cases, other examples of possible partitioning of the overall safety case can be seen in other standards. For example, the U.K. MoD Ship Safety Management Code JSP 430 in stating the following requirement for the production of safety cases clearly acknowledges the need for multiple safety cases for any single platform:

"Separate Safety Cases and Safety Case Reports are to be available for all units of the ship that are capable of independent operation, i.e. landing craft, boats, aircraft, weapon systems etc."

One might infer from this requirement that systems capable of independent operation implies the existence of independent safety cases. However, this will rarely be the case. Where systems are expected to operate as part of a coherent platform to provide a coherent capability, safety of any one element will often depend upon the safety of the other elements even if only because they define and share a common operating context and share common infrastructure (e.g. available human resource).

Outside of the defence context, the railway industry is another domain where the requirement for and existence of multiple interrelated safety cases is recognised. The Railway (Safety Case) Regulations 2000 (HSE, 2000) require that safety cases are submitted to the Health and Safety Executive (HSE) by railway operators (Infrastructure Controllers, Train Operating Companies, Station Operating Companies). The *implicit* overall safety case for safe U.K. train travel is made up of many safety cases for infrastructure, train operations and station operations. Safety cases for train operations will inevitably make assumptions of the infrastructure. Safety cases for station operators will inevitably make assumptions of the train operators. Boundaries may be drawn between these elements however interdependencies will always remain.

The European Railway Standard CENELEC 50129 (CENELEC, 1998) recognises the importance of recording the relationship between any single safety case any other safety cases to the extent that a section of the recommended safety case structure is reserved for this purpose. CENELEC 50129 talks of safety cases being structure into six parts:

- **Part One** – Definition of the System

- **Part Two** – Quality Management Report

- **Part Three** – Safety Management Report

- **Part Four** – Technical Safety Report

- *Part Five* – *Related Safety Cases*

- **Part Six** – Conclusions

Part Five of the safety case is given a dual role. Firstly, it should be used to record references to the safety cases of any subsystems or equipment on which the main safety case depends. Secondly, it should be used to present an account of the evidence of satisfying safety conditions from other safety cases. Perhaps because of the complexity of trying to provide a "joined up" safety case in the railway operations context, as described above, CENELEC 50129 has taken the very enlightened view that "no safety case is an island" by making these explicit recommendations for Part Five.

CENELEC 50129 also supports the principles of establishing multiple related safety cases in stating that the following three different types of safety case can be considered:

- *A generic product safety case* provides evidence that a generic product is safe in a variety of applications

- *A generic application safety case* provides evidence that a generic product is safe in a specific class of applications

- *A specific application safety case* that is relevant to one specific application

Within this framework the UK Railtrack Engineering Safety Management Handbook (Railtrack, 2000) (commonly known as the "Yellow Book") talks of possible re-use of safety evidence. It gives the example of a specific application Safety Case for a re-signalling scheme referring to a generic application safety case for the use of a points machine in a particular type of junction which may in turn refer to a generic product safety case for that points machine. This can be thought of as effectively adopting a modular safety case approach whereby new safety cases are made through the composition of existing safety cases with *new* arguments and evidence.

4 Emerging Classes of Systems

The previous section has sought to highlight that the requirements for, and principles of, a modular approach safety case construction are not new in so much as the approach is already required in reasoning about "conventional" systems. However, the need for a modular safety case approach is made even more apparent when considering some of today's emerging classes of systems, including the following:

- Off-The Shelf Components designed for use in the Safety Critical Sector
- "Systems of Systems"
- Integrated Modular Avionics

In the following sub-sections we will highlight how each is an example area where safety cases must be thought of in modular terms in order to reliably reason about whole system safety.

4.1 Safety Cases for COTS Marketed into the Safety Critical Sector

For a commercial vendor wishing to sell their component (e.g. a Real Time Operating System - RTOS) into the safety-critical marketplace *modules* there is an obvious commercial and marketing advantage for them to obtain some form of component certification. However, system is a safety property and they are only marketing a component that will eventually form *part* of a system. It is therefore infeasible for them to produce a self-contained safety case and any claim about overall system safety or integrity should be viewed with scepticism.

The current response of some vendors to this problem is to offer "certification packs" (usually for an additional fee!) which contains the raw evidence (often process based) from which any system integrator can attempt to construct a case as to why the component has sufficient integrity to be used in a particular application. This situation is analogous to a builder giving a house purchaser the raw building materials and asking *them* to build the house.

Many purchasers of COTS components supposedly designed specifically for the safety-critical sector want more than a bag full of evidence – they would like some form of *partial* safety case. In such a situation the component vendor will attempt to attempt to envisage the claims of their component that will typically be demanded in any system safety case. For example, in the case of an RTOS, it is easy to imagine that the system integrator will need to make claims regarding non-interference between process memory spaces or regarding the integrity of the scheduling policy and implementation. The component vendor will then assemble, as far as is reasonably practicable, the arguments and evidence required to support these claims.

The collection of component claims and supporting arguments and evidence do not make a complete safety case for the component. Rather they form a partially constructed safety case with arguments in a "ready to use" form within a target application safety case. With the house purchasing analogy, this would equate to a purchaser being able to buy prefabricated sections of their house ready to be assembled and completed within the context of a particular house design.

These partial safety cases can be thought of safety case modules, along the lines being suggested in this paper, and should be similarly managed. It is important to expose clearly the claims being supported by these cases, the context assumed and the evidence presented such that there can be an aware and intelligent application of this information in a system safety case context.

4.2 Safety Cases for "System of Systems (SoS)"

One class of systems increasingly being discussed in the safety critical sector (especially in relation to defence systems) are "System of Systems". The Oxford English Dictionary defines a (single) system in the following terms:

"An organised or connected group of objects

A set or assemblage of things connected, associated or interdependent so as to form a complex unity"

The description "System of Systems" (SoS) can therefore be used to refer to an assemblage of *systems* connected, associated or interdependent so as to form a complex unity. Examples of SoS include integrated tactical battlefield systems with land, sea and air systems coordinating to reach an overall objective (with obvious potential for unsafe interactions) and the implicit SoS established for safe control of European airspace by the coordinated action of multiple national Air Traffic Control agents. With such systems it is not possible to look at any single part and ask, "Is it safe"? The safety of the component part depends upon how it contributes to the safety of the whole the safety of the peer components with which it interacts.

Again, the safety cases for component systems in a SoS context require management as modular safety cases.

4.3 Integrated Modular Avionics

Traditionally aircraft computer systems have been *federated* – each system provided on a number of dedicated hardware units. The containment and physical isolation offered by federated systems has supported separate safety analysis and safety justification. However, the aviation industry is increasingly wishing to move towards Integrated Modular Avionics (IMA). Using the definition provided by Rushby (Rushby, 1999), in IMA a single computing system (with internal replication to provide fault tolerance) provides a common computing resource to several functions.

Integrated Modular Avionics (IMA) offer potential benefits of improved flexibility in function allocation, reduced development costs and improved maintainability, it poses significant problems in the development of the safety case. A principal motivation behind IMA is that there is through-life (and potentially run-time) flexibility in the system configuration. An IMA system can support many possible mappings of the functionality required to the underlying computing platform.

In constructing a safety case for IMA an attempt could be made to enumerate and justify all possible configurations within the architecture. However, this approach is unfeasibly expensive for all but a small number of processing units and functions. Another approach is to establish the safety case for a specific configuration within the architecture. However, this nullifies the benefit of flexibility in using an IMA solution and will necessitate the development of completely new safety cases for future modifications or additions to the architecture.

A more promising approach is to attempt to establish a modular, compositional, approach to constructing safety cases that has a correspondence with the modular structure of the underlying architecture. However, although the system architecture may be thought of as a neatly partitioned structure, safe overall system behaviour may not be so easily compartmentalized. IMA potentially suffers problems of poor fault containment. Owing to shared infrastructure (e.g. shared memory spaces, shared communications busses) the failures in one system element may propagate and cause failures in other system elements. Consequently, one must think very carefully about the structure of the overall safety case. Whilst parts of the safety case may directly correspond to system elements, other parts must explicitly address issues such as the integrity of the infrastructure and the intended and unintended interactions between elements. The need for a modular approach to safety case management in such cases is hopefully apparent. The architecture of the safety case must receive explicit and considerable attention.

5 'Hazards' in Safety Case Decomposition

The principle of decomposing system safety cases into subsystem safety cases may appear straightforward. However, there are a number of possible hazards in adopting this approach, including the following:

- **Failing to consider interactions between subsystem safety cases** – In the drive to "Divide and Conquer" the safety case we must be careful to ensure that at least one element of the safety case addresses the safety considerations posed by the interactions between elements.

- **Disproportionate effort allocated across overall safety case development** - It is important that a proportionate level of effort is dedicated to each subsystem safety case. It is unsatisfactory for one element of the overall safety case to be pursued in excess whilst others are under investigated or represented. When the safety case is taken as a whole a weak subsystem safety case may undermine other stronger dependent parts of the safety case.

- **Duplication of effort when apportioning responsibility** - This, of course, is not a safety problem but rather an efficiency problem. When dividing up the responsibilities of the subsystem safety cases care must be taken to clearly identify the scope and boundary of the subsystem safety case objectives.

- **Safety objectives "falling down the cracks" when apportioning responsibility** – This *is* a significant safety issue. It is all too easy when dividing up responsibilities among subsystem safety cases for one sub safety case to assume that another is tackling an objective and vice versa. Obviously, if unnoticed, an incomplete safety case will result.

An underlying cause of these problems is a poor understanding of the overall structure of the system safety case and safety argument. By explicitly planning (at an early stage in the lifecycle), representing, and managing the safety case architecture we can hope to alleviate some of these problems.

6 Safety Case Architecture Styles

In the field of software architecture there are a number of recognised architectural 'styles'[1]. These styles define patterns of component types and possible interactions and include examples such as "Data-Centric", "Data Flow", "Event-Driven" styles (Bass et al., 1998). They describe fundamental principles and constraints governing the organisation of the software architecture. In the same way it is possible to describe styles in the organisation of safety case architecture.

The simplest safety case architecture style is for the safety case decomposition into sub-safety cases to attempt to follow (as far as is possible) the system decomposition. This approach is suggested within Annex E of Part 2 of Defence Standard 00-55:

> "*E.2.4.2 Production of the Software Safety Case should follow the approach adopted for the System Safety Case. For example, one possible approach for a complex system might involve developing a top-level System Safety Case which is supported by lower level Subsystem Safety Cases. Each Subsystem which contains SRS should then have its own Software Safety Case. It may also be useful to prepare a top-level Software Safety Case which covers common aspects of all software in the system.*"

This form of safety case division may be particularly appropriate when system components are being developed to different levels of integrity or assurance. This is also recognised by 00-55:

> "*{A} Software Safety Case for a large software system composed of large subsystems of differing safety integrity levels might be structured as:*
>
> *(a) a set of Subsystem Software Safety Cases justifying the safety of each software subsystem;*
>
> *(b) a top-level Software Safety Case justifying how the combination of subsystems is safe.*"

As described earlier in section 4.3, this approach whilst being possible for conventional (statically defined) software systems would prove infeasible for highly configurable and adaptable modular systems such as IMA. Rather than reasoning about all possible combinations of systems (of which there will be many!) we require a safety case architecture exhibiting one or both of the following styles:

- **Reasoning about configuration blueprints** – A set of blueprints of possible system combinations, ideally covering *classes* of combinations is

---

[1] Architectural 'Styles' are synonymous with Architectural 'Patterns'

identified for the IMA system a priori. The safety case architecture, in addition to having a conventional subsystem safety case structure would also have a safety case element (or 'module') that argues the safety of the blueprint definition. Such a safety case architecture permits safe operation of any system configuration that can be said to fall within the system blueprint. Blueprints are discussed further in section 8.

- **A 'Backplane' Safety Argument** – the extent of required reasoning about possible interactions between system components can be limited if it is possible to establish a sound 'backplane' argument concerning the infrastructure of the modular system. Particularly, such an argument must support claims regarding the system partitioning. For example, if it is possible to establish an argument (for the infrastructure) that memory spaces are properly segregated and that possible interference via communications busses is predictable and manageable then this limits the extent to which *each* subsystem safety case (e.g. for each avionics application) must reason about such issues.

Rather than decomposition of safety case according to system structure another 'style' is to decompose the case according to safety *functions*. Functional division may cut across system (and therefore organisational) boundaries. Multiple systems may contribute to the performance of a particular safety function. Similarly, systems may contribute to the performance of more than one safety function. This style has one advantage over the subsystem decomposition style in that it promises to be more cohesive *from a safety perspective*. For someone wishing to look at all of the issues surrounding the performance of a particular safety function they will find them largely addressed within a single sub safety case. However, in systems that where safety functions are heavily integrated we run the risk of failing to address the safety issues associated with function interaction. To combat this, it is essential to ensure that either function interactions are considered within each sub function safety case (*good* for comprehensibility but *bad* for maintainability) or within a separate 'interaction' sub safety case (*less* comprehensible but *easier* to maintain).

In the same sprit as the functional decomposition style, it is also possible to decompose the safety case according the structure of identified hazards. This has obvious advantages for aiding understanding of safety. However, hazards cut across system and function boundaries and unless the work division and organizational structure for producing the safety cases can be divided upon along those lines this approach may be hard to realise in practice. In effect, hazard IPTs (Integrated Project Teams) would need to be created to support this approach. As with functional decomposition, we also need to ensure that interactions (in this case between hazards) are addressed somewhere within the overall structure of the safety case.

The styles described above are only a few of many potential ways of organizing the safety case structure "in-the-large". Other examples include division across process argument and product argument boundaries, and organization according to styles of evidence (e.g. having Deterministic safety case elements distinct from Probabilistic safety case elements). As people become more familiar with, and practiced in, reasoning about safety case architecture more architectural styles will naturally emerge.

7 Representing Modular Safety Cases

The Goal Structuring Notation (GSN) (Kelly, 1997) - a graphical argumentation notation - explicitly represents the individual elements of any safety argument (requirements, claims, evidence and context) and (perhaps more significantly) the relationships that exist between these elements (i.e. how individual requirements are supported by specific claims, how claims are supported by evidence and the assumed context that is defined for the argument). The principal symbols of the notation are shown in Figure 1 (with example instances of each concept).

The principal purpose of a goal structure is to show how goals (claims about the system) are successively broken down into sub-goals until a point is reached where claims can be supported by direct reference to available evidence (solutions). As part of this decomposition, using the GSN it is also possible to make clear the argument strategies adopted (e.g. adopting a quantitative or qualitative approach), the rationale for the approach (assumptions, justifications) and the context in which goals are stated (e.g. the system scope or the assumed operational role). For further details on GSN see (Kelly, 1997).

Figure 1 – Principal Elements of the Goal Structuring Notation

GSN has been widely adopted by safety-critical industries for the presentation of safety arguments within safety cases. However, to date GSN has largely been used for arguments that can be defined 'stand-alone' as a single artefact rather than as a series of modularised interconnected arguments. In order to make the GSN support the concepts of modular safety case construction it has been necessary to make a number of extensions to the core notation.

The first extension to GSN is an explicit representation of modules themselves. This is required, for example, in order to be able to represent a module as providing the solution for a goal. For this purpose, the package notation from the Unified Modelling Language (UML) standard has been adopted. The new GSN symbol for safety case module is shown in Figure 2 (Right Hand Side).

As has already discussed, in presenting a modularised argument it is necessary to be able to refer to goals (claims) satisfied within other modules. Figure 2 (left hand side) introduces a new element to the GSN for this purpose – the "Away Goal". An away goal is a goal that is not supported within the module where it is presented but is instead supported in another module. The Module Identifier (shown at the bottom of the away goal next to the module symbol) should show the unique reference to the module where support for the goal can be found.

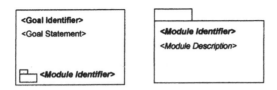

Figure 2 – GSN Elements Introduced to Handle Modularity

Away goals can be used to provide *support* for the argument within a module, e.g. supporting a goal or supporting an argument strategy. Away goals can also be used to provide contextual backing for goals, strategies and solutions.

Representation of away goals and modules within a safety argument is illustrated within Figure 3. The annotation of the top goal within this figure "SysAccSafe" with a module icon in the top right corner of the goal box denotes that this is a 'public' goal that would be visible as part of the published interface for the entire argument shown in Figure 3 as one of the "objectives addressed".

The strategy presented within Figure 3 to address the top goal "SysAccSafe" is to argue the safety of each individual safety-related function in turn, as shown in the decomposed goals "FnASafe", "FnBSafe" and "FnCSafe". Underlying the viability of this strategy is the assumed claim that all the system functions are independent. However, this argument is not expanded within this "module" of argument. Instead, the strategy makes reference to this claim being addressed within another module called "IndependenceArg" – as shown at the bottom of the goal symbol. This form of reference to a goal being addressed within another (named) module is called an "Away Goal". The claim "FnASafe" is similarly not expanded within this module of argument. Instead, the structure shows the goal being supported by another argument module called "FnAArgument". The "FnBSafe" claim is similarly shown to be supported by means of an Away Goal reference to the "FnBArgument" module. The final claim, "FnCSafe", remains undeveloped (and therefore requiring support) – as denoted by the diamond attached to the bottom of the goal.

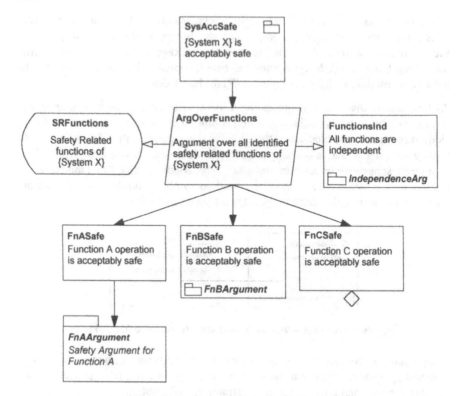

Figure 3 – Representing Safety Case Modules and Module References in GSN

In the same way that in can it be useful to represent the aggregated dependencies between software modules in order to gain an appreciation of how modules interrelate "in-the-large" (e.g. as described in the "Module View" of Software Architecture proposed by Hofmeister et al. in (Hofmeister et al., 1999)) it can also be useful to express a module view between safety case modules.

If the argument presented within Figure 3 was packaged as the "TopLevelArg" Module, Figure 4 represents the module view that can be used to summarise the dependencies that exist between modules. Because the "FnAArgument" and "FnBArgument" modules are used to support claims within the "TopLevelArg" module a supporting role is communicated. Because the "IndependenceArg" module supports a claim assumed as context to the arguments presented in "TopLevelArg" a contextual link between these modules is shown.

In a safety case module view, such as that illustrated in Figure 4, it is important to recognise that the presence of a *SolvedBy* relationship between modules A & B implies that there exists at least goal within module A that is supported by an argument within module B. Similarly, the existence of an *InContextOf* relationship between modules A & B implies that there exists at least one contextual reference within module A to an element of the argument within module B.

Alongside these extensions to the graphical notation of GSN, the following items of supporting documentation are required:

- **Interface declaration for each safety case module** – along the lines outlined in section 2, the external visible properties of any safety case module must be recorded – e.g. the goals it supports, the evidence (solutions) it presents, the cross-references ('Away Goal' references) made to / dependencies upon other modules of argument.
- **Contracts for composed modules** – where co-dependent safety case modules are used together within a system safety case a contract must be recorded of the dependencies resolved between the separate arguments.

Due to space limitations it is not possible to describe details of safety case module interface definitions or contracts. For further details see (Kelly, 2001).

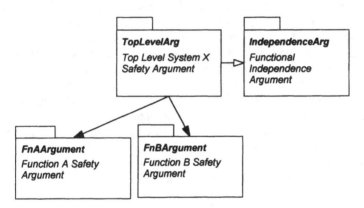

Figure 4 – Example Safety Argument Module View

8 Implications for Certification Processes

A modular approach to safety case construction has implications on the certification and acceptance processes. Whereas, traditionally certification has involved accepting, at a single point in time, a single safety case for an entire system for the benefits of a modular safety case approach to be realised requires a certification process that acknowledges the structure of a partitioned safety case that can be extended and modified without instantly requiring re-evaluation of the entire case. The guidance document ARINC 651 (ARINC, 1991) recognises this fact for suggests that for IMA-based systems the certification tasks are comprised of the following three distinct efforts:

- Confirmation of the general environment provided by the cabinet
- Confirmation of the operational behaviour of each function (application) intended to reside within a cabinet
- Confirmation of the resultant composite of the functions

ARINC 651 also recognises that conventional safety standards (such as DO178B (RTCA, 1992)) may need to be updated to reflect these new distinct tasks. ARINC 651 also talks explicitly of the need for "building block qualification" whereby it is possible to "separately quality certain building blocks of an IMA architecture in order to reduce the certification effort required for any particular IMA-hosted function". Example building blocks listed include specific arguments relating to the (ARINC 629) global data bus, the ARINC 659 backplane bus, the robust partitioning environment and the cabinet hardware / software environment. However, no detail regarding how these building block arguments are to be represented and managed is presented within ARINC 651.

In order to design and validate the various building blocks involved in IMA, ARINC 651 identifies the need for "rules which govern how the building blocks work together". It additionally describes that, "a feature of these rules of application is that they can be used to limit the work associated with certifying and re-certifying an IMA function to proof of compliance with the rules, and qualification of the function itself. Regulatory agency discussion is encouraged to establish how certification credit may be granted for adherence to these rules". This concept of defining rules between building blocks relates strongly to the principles of establishing well-defined module interfaces and contracts between safety case modules put forward within this paper. As the quote above clearly highlights, a necessary part of a new certification process based upon modular safety cases is to clearly give credit (i.e. limit the required re-certification) where contracts between safety case modules are upheld in the light of change to, or reconfiguration of, modules within the overall safety case.

9   Summary

In this paper we have described a key mechanism for managing safety case complexity – the concept of *safety case architecture* and (consequently) *modular safety cases*. Whilst the idea might appear new we have shown how it relates clearly to existing concepts of constructing whole system safety cases from a number of sub-system safety cases as described in current safety standards. To manage a modular safety case development *safely* we must be explicit about the interfaces of safety case modules, and acknowledge and represent clearly the dependencies that exist between modules. Extensions to the Goal Structuring Notation (GSN) for this purpose have been described.

Safety is not a "sum of parts" property. Care must therefore be taken in how safety cases are divided up such that interactions are recognised and addressed within the safety argument. Example hazards posed by safety case decomposition and styles of decomposition (safety case architectural 'styles') have been discussed. Finally, we have highlighted the possible implications for certification processes of adopting a modular safety case approach.

10 Acknowledgements

The author would like to acknowledge the financial support given by QinetiQ for some of work reported in this paper.

11 References

ARINC (1991) Design Guidance for Integrated Modular Avionics, Aeronautical Radio, Inc.

Bass, L., Clements, P. and Kazman, R. (1998) *Software Architecture in Practice,* Addison-Wesley.

CENELEC (1998) *ENV 50129 Railway applications - Safety related electronic systems for signalling,* European Committee for Electrotechnical Standardisation.

Hofmeister, C., Nord, R. and Soni, D. (1999) *Applied Software Architecture,* Addison-Wesley.

HSE (2000) *Railway Safety Cases - Railway (Safety Case) Regulations 2000 - Guidance on Regulations,* HSE Books.

Kelly, T. (2001) Concepts and Principles of Compositional Safety Case Construction (Contract Research Report for QinetiQ COMSA/2001/1/1), Department of Computer Science, University of York (available from www.cs.york.ac.uk/~tpk/pubs.htm)

Kelly, T. P. (1997) *A Six-Step Method for the Development of Goal Structures,* York Software Engineering.

Kelly, T. P. and McDermid, J. A. (2001) A Systematic Approach to Safety Case Maintenance, *Reliability Engineering and System Safety,* **71,** 271.

Meyer, B. (1992) Applying Design by Contract, *IEEE Computer,* **25,** 40-52.

MoD (1996) *Defence Standard 00-56 Safety Management Requirements for Defence Systems,* Ministry of Defence.

MoD (1997) *Defence Standard 00-55, Requirements of Safety Related Software in Defence Equipment,* Ministry of Defence.

Railtrack (2000) Engineering Safety Management - Issue 3, Electrical Engineering and Control Systems, Railtrack

RTCA (1992) Software Considerations in Airborne Systems and Equipment Certification, RTCA

Rushby, J. (1999) *Partitioning in Avionics Architectures: Requirements, Mechanisms, and Assurance (NASA Contractor Report CR-1999-209347),* NASA Langley Research Center.

DEVELOPMENT AND LEGAL ISSUES

DEVELOPMENT AND LEGAL ISSUES

White Box Software Development

Dewi Daniels, Richard Myers and Adrian Hilton
Praxis Critical Systems
20 Manvers Street, Bath BA1 1PX, England

Abstract

This article attempts to debunk the populist view that building high quality software is difficult and costly, and that having software systems that crash is an acceptable state of affairs.

The technology to build predictable reliable software systems exists today. Principled engineering judgment can be used to tailor software development so that quality can be built in with cost in mind – this is particularly the case with safety critical systems, where the application of standards can force an unnecessarily rigorous approach for little proven benefit.

This article explores the general poor quality of software "in the large", the public's (and the industry's) view that this is in some way acceptable, and then presents some real case studies which show how quality can be built in without the need to invest in overweight tools and technologies.

1 Introduction

The average customer of the computing industry has been served so poorly that he expects his system to crash all the time, and we witness a massive world-wide distribution of bug-ridden software for which we should be deeply ashamed.

Edsger Dijkstra, Communications of the ACM, vol. 44 no. 3, March 2001

1.1 Poor Quality Is The Norm

Software is generally sold with little or no meaningful warranty. The media is often the only part of a package with a guarantee. The contrast between the end-user licence agreements for software and hardware is shown in the program licence agreement from Compaq (Compaq 1999):

Compaq Computer Corporation warrants the media on which the programs are furnished, to be free from defects in materials and workmanship under normal use for a period of one year from the date of delivery to you [...]

YOU ASSUME THE ENTIRE RISK AS TO THE QUALITY, USE AND PERFORMANCE OF THE PROGRAMS. SHOULD THE PROGRAMS

PROVE DEFECTIVE, YOU – AND NOT COMPAQ OR ITS SUPPLIERS OR AN AUTHORISED RESELLER – ASSUME THE ENTIRE COST OF NECESSARY SERVICING, REPAIR OR CORRECTION.

1.2　The Industry Has Accepted This Situation

Software projects are frequently too late, too expensive and unable to meet the customer's real needs. In the three-way trade-off between time, cost and functionality, software often fails to deliver any of the three.

The natural result of this is that the industry no longer believes that it is possible to deliver adequate software on time and to budget. The consequence is a widespread practice of attempting to improve software quality solely by finding and fixing faults in finished artefacts, whether units or systems.

> *I've said before that the best way to find bugs is to execute the code and then somehow spot them...*
> Steve Maguire, "Writing Solid Code", Chapter 4, Microsoft Press, 1993

1.3　White Box Software Is Different

We might call this generate, test and rework paradigm "black-box software development" because it is fundamentally unconcerned with the internals of the software development process. Building-blocks are assembled or re-assembled, and getting it out of the door is all that matters. A "one size fits all" approach is taken to both the product and process; no assessment is made of whether some parts of the product may be better served by a different process.

By contrast, an approach in which specific attention is paid to the product being built and the development process being used might be dubbed "white-box". This approach tailors the process and methods used to the success criteria of the project and the heterogeneity of the product.

1.4　Structure

In this paper we review the current state of the industry, and then argue that the low expectations of software are the result of three fallacies:

- that software is inherently complex and error-prone;
- that tools or methods will fix this problem; and
- that Software Engineering cannot be a true engineering discipline.

We then show that it *is* possible to produce high quality software on time and on budget by applying sound engineering principles throughout the life-cycle. There is no single method for doing this – there is still no silver bullet (Brooks 1995). However there are common themes:

- deploy small, highly-skilled teams;
- use tailored, focused processes; and
- choose the appropriate tools for the job.

2 State of The Practice

It's hard to read through a book on the principles of magic without glancing at the cover periodically to make sure it isn't a book on software design.

Bruce Tognazzini, "Principles, Techniques and Ethics of Stage Magic and their Application to Human Interface Design", Proceedings of INTERCHI, April 1993

2.1 The Costs of Failure Are High

A recent report from the American National Institute of Standards and Technology suggests that software faults cost the US economy some $40bn to $60bn annually (Tassey 2002). The costs to the European Union will be of a similar order. About 60% of these costs are borne by the customer. The rest are borne by the vendor. The report surveyed software developers and users in the automotive, aerospace and financial services sectors. The cost estimates did not include "mission critical" software.

2.2 The Causes of Failure Are Well Understood

Most IT projects fail (i.e. are cancelled before completion) because problems which should have been dealt with at early stages are not detected until it is too late to correct them. The common causes of IT project failure are revealed by two studies. The first was conducted by the Standish Group in the USA (Standish 1995). The second, smaller, study was conducted by Taylor in the UK (Taylor 2001). The results are summarised in Figure 1.

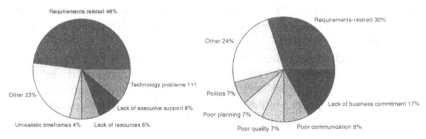

Figure 1. Undetected faults in early life-cycle stages cause project failure.
The reported causes of failure are most often requirements-related and yet these projects often fail after considerable time and expense. The truth is that requirements problems are left undetected until late in the life-cycle at which point they cannot be rectified.
Based on data from (a) (Standish 1995) and (b) (Taylor 2001).

The 1994 Standish Group CHAOS report found that:

- Some 31% of projects failed at a cost of $81bn.
- An additional 53% were troubled.
- Only 16% were successful.

The mean time and budgetary overruns were both around 200%, with just over half the original functionality actually delivered. Almost all of the perceived causes were upstream of implementation.

Taylor's study was based on detailed questioning of 38 members of the British Computer Society, The Association of Project Managers and The Institute of Management. Of 1027 projects studied, only 130 were successful, of which:

- 2.3% were development projects;
- 18.2% were maintenance projects; and
- 79.5% were data conversion projects.

Out of over 500 development projects in the sample, only 3 were successful. About half of the projects which failed were cancelled prior to implementation.

A recent update to the CHAOS report (Standish 1999) indicates that by 1996 the proportion of successful projects had increased to 27%, but that this had stayed the same up to 1998, suggesting that the project success rate had reached some form of limit which left 3 out of 4 projects still not delivered to acceptable levels of quality within timescale and budgetary constraints.

2.3 But The Failures Keep Coming

2.3.1 Delivered Systems

Expensive software errors in delivered systems are not new. The first famous case was the loss of the Mariner 1 probe to Venus in 1962 as a result of an error in a FORTRAN statement: the programmer had typed a "." instead of a ",".[1]

In the mid-1980s at least six people were exposed to massive doses of radiation as a result of race conditions in concurrent software for the Therac-25 radiation therapy machine (Leveson 1995). The software had been re-used from an earlier model, but the errors had not been detected because the earlier model had hardware interlocks to prevent overdose.

In 1991, 28 soldiers died at Dhahran following the failure of a Patriot missile to intercept an incoming Scud. The software had never been designed to run continuously for long periods and had a cumulative rounding error in it. At Dhahran the Patriot system had been left on for 100 hours and the cumulative error was enough to cause the tracking system to fail.

---

[1] The use of a programming language where a single-character error can change the semantics of the program is clearly inappropriate for such a high-cost enterprise. In 1962 they probably had nothing better. However, today we find C and C++ used in a variety of mission-critical settings, where confusing = and == may have similar results.

In 1996 the Ariane 501 launcher was lost with payload at an estimated cost of $1bn. The cause of the loss was a run-time exception in a software unit which had been re-used from Ariane 4 but was, ironically, not needed in Ariane 5 (Lions 1996).

Most recently, in 1999 the $130m Mars Climate Orbiter collided with Mars due to an embarrassing confusion between metric and imperial units (Stephenson et al. 1999).

The apparent simplicity of these errors belies the complexity of the underlying causes. None of these accidents could truly be described as the result of a simple programming error, although software failure was the proximate cause of each.

2.3.2   Cancelled and Late Projects

Cancelled projects, or those which are delivered late at excessive cost, are more significant in terms of cost because they are more common. One example was the 1990 Wessex Health Authority Regional Information Systems Plan, which cost the Authority £43m (Committee of Public Accounts 1993) before being abandoned. Poor project management exacerbated the result of the software failings, perhaps by up to £20m.

The Bowman project to provide the British Army with a secure communications system has suffered over 75 months of delays and went £200m over budget (Hansard 2000). The lack of this capability could leave forces vulnerable to electronic counter-measures and unable to use secure lines of communication.

2.4   And We Just Accept It

The most outrageous aspect of this situation is that it has been accepted as the industry norm. The difficulty of producing a high-quality product (and it is difficult to engineer good software) should inspire us to strive harder rather than just give up. Unfortunately the latter is what appears to have happened. We believe that tolerance of this sorry state is the result of the three widespread fallacies which we discuss in the next section.

3   Three Widespread Fallacies about Software Development

This section describes and refutes the three fallacies which we believe lead to unjustifiably low expectations of software development:

- that software is inherently complex and error-prone;
- that tools or methods will fix the problem; and
- that Software Engineering is not a true engineering discipline.

3.1 Fallacy: Software Is Inherently Complex

There are two ways of constructing a software design. One way is to make it so simple that there are obviously no deficiencies. And the other is to make it so complicated that there are no obvious deficiencies.

Tony Hoare, ACM Turing Award Lecture, 1980

The first fallacy is that we should expect software to be unreliable because it is complicated. The truth is that software is complex mainly because we choose to make it so. The industry has become used to creating complex software solutions where simple ones would do. Complex software arises because:

- software construction begins prematurely;
- it is easy to achieve partial success at the expense of complexity; and
- writing simple software is difficult.

Premature commitment to software creation is a pervasive problem in the industry. The pressure to be seen to be making progress, coupled with the desire to get coding means that software artefacts are created before the problems they address are adequately understood (Leveson 1995, McConnell 1999). This tendency contributes to the large number of failures attributed to "requirements-related" causes. If software has been created before the requirements are well understood, requirements changes will be difficult to accommodate and the final product will be expensive, late, incomplete and/or defective.

It is easy to create complex software which works most of the time, but this is at the expense of unmanageable complexity which renders further improvement impossible (Leveson 1995). Repeated updates to software to fix bugs will tend to increase the software's complexity (the principle of "software entropy"), eventually to a point where the software is too complex to change without breaking existing functionality. Such software is often termed "fragile".

The difficulty of writing simple software is not immediately obvious. In the physical world, the properties of raw materials provide natural constraints on modifications. Building a physical machine involves purchasing and assembling components and possibly tooling-up a factory. There are few such obvious constraints on software: changing it appears to be a simple matter of rewriting some text. The traditional constraints of memory, processor speed and turnaround time no longer apply, so programmers are free to create code in an ad-hoc manner. The only requirement is that the code compiles.

Such an approach ensures that only the software's author can understand or fix it, which carries a certain attraction for some programmers. The fact that *no-one* understands the code becomes apparent as attempts to fix the remaining defects prove futile. Unfortunately for the customer, the "heroes" who create dazzlingly complex software so quickly and then fight so valiantly to fix the bugs are often better rewarded than those who take the time to get things right.

The result of the fallacy of complexity is that customers have come to regard rapid fixes as the pinnacle of quality; not introducing the fault in the first place is less well regarded (Stålhane et al. 1997, Chulani et al. 2001). In fact, in software development as in so many other activities, it is the mark of a skilful practitioner that they make things look simple. Elegance and simplicity are hard-won but the end result looks trivial to the inexperienced eye.

3.2 Fallacy: Tools and Methods Will Fix The Problem

It is impossible to sharpen a pencil with a blunt axe. It is equally vain to try to do it with ten blunt axes instead.
Edsger W. Dijkstra, "How do we tell truths that might hurt?", June 1975

The second fallacy is that tools and methods alone will improve software quality. We are not arguing that tools and methods are useless, but that they are no substitute for skill and experience. Software tools and methods are often expensive to deploy and require specialist knowledge to use and support. The inappropriate use of a method may hinder rather than assist a project. Several of our clients in aerospace, rail and telecommunications have experienced significant difficulties stemming from the inappropriate use of methods and tools.

3.2.1 No Silver Bullet

Brooks's 1986 prediction of "no silver bullet", that no single Software Engineering development would yield an order of magnitude improvement in programming productivity in ten years, was borne out by events (Brooks 1986, Brooks 1995). Indeed, many commentators questioned whether all the developments taken together over that period had led to an order of magnitude improvement.

No language is a substitute for programming skill. Similarly, no requirements, design or testing tool can eliminate the need for engineering skill and experience.

3.2.2 Everything Old Is New Again

Many of the significant "developments" of the 1980s and 1990s were not actually new. Object oriented programming was first introduced in 1967, and program proof was first mooted in 1946. Many of the project life-cycle models used today date back to the 1970s. Fashions may change but fundamentals do not.

Worse, modern re-inventions of old concepts may not incorporate the lessons which were painfully learned by the adopters of the original idea. An example is the confusion between *abstraction*, which excludes unnecessary detail, and *information hiding*, which is the removal of state from visibility. The latter has been the focus of object-oriented programming, whereas the former should have been the focus (Amey 2001).

The effect of information hiding is that state and the way in which it changes can become complicated without the programmer realising. Abstraction makes things less complicated by hiding unnecessary detail, but information hiding increases complexity by brushing it under the carpet. Devising a useful abstraction is a highly skilled activity; stuffing state into the private part of a class is not.

3.2.3 Hidden Costs

Software tools may appear to save effort at a particular life-cycle stage, but this may be at the expense of other life-cycle stages. For example, automatic code generators may reduce the effort required in the coding phase. However, the effort required for design may increase correspondingly because the designer must provide additional information to direct the code generation.

Code generators may also have a negative impact on the effort needed for verification and validation. Since this is often 60-70% of the total effort, code generation tools should be judged by how much they simplify verification and validation rather than by how much coding time they save.

The idea that tools somehow encapsulate prior knowledge is a common but dangerous misconception. Its implication is that the software process can be de-skilled and effectively turned into a production line staffed by large teams of relatively unskilled technicians. Programmers did not become less skilled with the advent of compilers. Compilers enable programmers to shift their attention from the machine domain to the problem domain where it is needed. The inappropriate use of tools causes attention to shift back from the problem domain to the tool domain.

3.3 Fallacy: Software Engineering Cannot Be A True Engineering Discipline

> *Today we tend to go on for years, with tremendous investments to find that the system, which was not well understood to start with, does not work as anticipated. We build systems like the Wright brothers built aeroplanes – build the whole thing, push it off the cliff, let it crash, and start over again.*
>
> Professor R. M. Graham, Massachusetts Institute of Technology, from the debate "Software Engineering and Society", NATO Science Committee conference on Software Engineering, October 1968

The third fallacy is that Software Engineering cannot be a true engineering discipline. It is certainly arguable that current practice in software development would not be acceptable in many engineering disciplines. Our position is that current practice should be unacceptable in the software industry also.

This fallacy is often expressed as "programming is an art, not a science". However, there is much more to Software Engineering than programming. The fact that very few project failures are directly attributed to bad programming and relatively few to poor project management suggests that the problem lies in between these two activities. That is the place of Software Engineering.

There is a substantial body of knowledge in software engineering, dating back over thirty years. The first conference on the subject was in 1968, and the above quotation from that conference is still true thirty four years later.

3.3.1 Software Engineering vs. Programming

Three examples distinguish Software Engineering from programming. The first is the 1996 Ariane 501 failure (Lions 1996). The immediate cause was a run-time exception generated by a piece of legacy software from Ariane 4. The software function was not actually needed in Ariane 5 and could therefore have been removed.

The second example is the loss of the Mars Climate Orbiter in 1999 due to a data file containing imperial rather than metric data (Stephenson et al. 1999). Both of these expensive and very embarrassing losses stemmed from software failures which were nevertheless *not* simple programming errors. They were, respectively, the failure to understand the impact of a domain change and the failure to understand an interface.

The third example is the Therac-25 overdosing failure (Leveson 1995). Here, the designer used software to maintain safety, instead of using the hardware interlocks present in the older Therac-20 machine; a software fault that killed and injured a number of patients treated by Therac-25 was also present in Therac-20, but the latter's hardware interlocks prevented any patient injury. This was a failure of *design* by the Therac-25 builders.

3.3.2 The Discipline of Software Engineering

Both the academic and professional worlds accept Software Engineering as a discipline. Many universities throughout the world offer undergraduate and postgraduate degrees in Software Engineering. Certification and licensing of software engineers is under study in the USA, notably in Texas and Illinois. Canada licenses software engineers, and the Canadian Council of Professional Engineers has advised Microsoft that unlicensed MCSE and MCPSE holders who call themselves "engineers" could face enforcement measures. The Engineering Council of the UK recognises that software engineers may be awarded Chartered Engineer status after appropriate training and experience.

We believe that software development processes are generally more mature than digital electronic design processes. Many functions which would once have required custom hardware are now implemented by programmed devices. The circuit design is expressed in VHDL source code. Leveson's "curse of flexibility" (Leveson 1995) now applies to hardware as well as to software. We have found on a number of recent projects that the software development has proceeded more smoothly than the hardware development. On one project, the software development was completed within a few months but hardware problems delayed the project by nearly a year.

We have found that projects in more traditional engineering enterprises such as the railway industry suffer from requirements-related problems. It has been our experience that, far from not being a true engineering discipline, Software Engineering has lessons to teach other disciplines.

3.4 Summary

These fallacies about software development are understandable, but demonstrably incorrect. Software need not be complex and unreliable; however, to make it reliable tools and methods are not enough. We need to treat Software Engineering as the engineering discipline that it should be.

Now that we have examined and refuted these fallacies we provide practical suggestions for improving software quality, based on our own experience of software development projects.

4 White Box Software Development

The advantage of being correct is that you do not need to change your mind.

J. K. Galbraith, BBC interview, 1996

Our fundamental philosophy is that it is quite possible to produce high-quality software. Our experience has been that producing software with very few defects is not only possible, but highly desirable: we spend less time and money fixing faults. As a result, our systems have low defect rates and deliver the required functionality on time and to budget. We achieve this by:

- deploying small, highly-skilled teams;
- using tailored, focused processes; and
- choosing the right tools for the job.

4.1 Deploy Small, Highly-Skilled Teams

I'm very impressed by the calibre of your people.

Andy Calvert, Security Development Manager, MONDEX International

Agencies frequently send out lists of ... skills [such] as Visual Basic, C++ etc. These are not skills, they are programming languages... As an aeronautical engineer I am amused by the idea of applying for a job at Boeing ... quoting my "skills" as: screwdriver, metric open-ended spanners and medium-sized hammers!

Peter Amey, "Logic versus Magic in Critical Systems", keynote address at Ada Europe, May 2001

For example, we employ about 100 engineers of whom 60% are software engineers; the others are safety and systems engineers. All of our engineers are degree-educated and many hold post-graduate qualifications; 25% hold doctorates and another 25% hold masters degrees. We have been delivering software successfully across the certification range for 20 years.

Our engineers are actively encouraged to achieve national and international recognition through gaining Chartered Engineer status with a recognised industrial body, for example the British Computer Society or the Institution of Electrical Engineers. A third of our engineers have Chartered Engineer status.

As professional engineers our staff continually seek to learn from others' experience, as well as improving wider industry practice through sharing our experiences. Paradoxically, it is because we recognise that engineering skill and experience are more valuable than intimate knowledge of the latest fad that we have knowledge of a variety of methods, tools and programming languages.

4.2   Use Tailored, Focused Processes

If you do not actively attack the risks, they will attack you.
Tom Gilb, "Principles of Software Engineering Management", pub. Addison Wesley, 1988

The key tenet of White Box Software Development is to tailor the development process to fit the project. We do not mandate any particular process model, requirements engineering method, specification technique, programming language or test and integration strategy. Rather we select and tailor processes and tools depending on the particular job.

We use risk-based planning (Ould 1999) for every project. We start by identifying the project risks in order to determine what process model to adopt and what risk reduction measures to put in place. We identify the key attributes of the project and client expectations. This enables us to define or select appropriate processes, methods and tools.

Believing that it is more efficient to prevent errors rather than try to detect and fix them, we have developed the REVEAL method for requirements engineering. REVEAL is based on sound engineering principles, a practical and clear method for eliciting and defining requirements; for improving specification clarity; and for managing change to those requirements.

The result is that we produce software with very low defect rates as shown in Figure 2, but without having to compromise on features, time or cost.

4.3   Choose The Right Tools for The Job

To a man with a hammer, everything looks like a nail.
Mark Twain

We do not start a project with any particular set of methods and tools in mind. We defer such decisions until we have defined the technical approach. This means that not only do we select the most appropriate methods and tools, but also that we are free to use those methods and tools in ways that maximise benefits to the client.

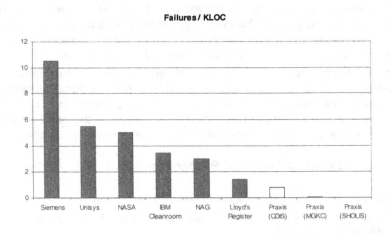

Failures / KLOC

Figure 2. Our post-delivery defect rates are very low.
The use of tailored processes allows us to focus on the real problems. The result is very low defect rates without compromising delivery. Source data from (Pfleeger & Hatton 1997).

For example, suppose our risk-based planning activity requires us to show that a piece of software meets certain requirements. Depending on the level of assurance needed, we would choose an approach somewhere on the spectrum from:

- simply tracing the software units to the requirements; to
- formally specifying the requirements, proving that the formal specification is consistent, writing the code in a high-integrity language, and then proving that the code met the specification.

We might choose a formal language like Z or VDM to specify a whole system or just a critical sub-system, depending on the circumstances. We would then decide whether or not to prove the specification. It may well be that most of the benefit is gained from the act of writing the specification. For some applications, full proof might be essential; for others it might be gold-plating.

Where our chosen programming language is SPARK Ada, we still have degrees of flexibility in its use. We may only use it as a design language and employ the Examiner static analysis tool to check information flow between design units; we may write the software in the SPARK subset and choose whether or not to do information flow analysis; we may attempt to prove the code free of run-time exceptions, or we may undertake partial proof of correctness of selected parts of the program. These choices will be made according to the results of our planning.

We might use SPARK Ada rather than C++ as the implementation language for a safety-critical controller, but C++ rather than SPARK Ada for a non critical GUI.

Our decision would be informed by expertise in *both* languages, rather than received ideas or prejudices.

5 Examples

This section contains recent examples of our work that illustrate the benefits of our White Box approach. With the exception of CDIS, all of these projects have been completed within the last 5 years.

5.1 Example: 10-Year Warranty and No Claims

A key point in choosing Praxis was that we respected ... their software engineering processes and felt confident that they would be able to deliver the system. We regard our choice as vindicated by the trouble-free manner in which the system has been integrated.

Derek McLauchlan, CEO responsible for the CAA Central Control Function programme

In the early 1990s, Praxis developed the SIL-2 Central Control Function Information Display System (CDIS) for the then Civil Aviation Authority, now National Air Traffic Services (Hall 1996). CDIS forms a vital component of the data entry and display equipment used by air traffic controllers at the London Air Traffic Control Centre (LATCC). The key characteristics of CDIS are:

- Very high availability (24/7) distributed real-time system
- Risk-based planning
- Prototyping and interviews to gather requirements
- Formal specifications using VDM and CSP
- Proof of critical parts only
- Implemented in C (200,000 lines)
- 0.81 defects per KLOC
- 10 year warranty
- No claims yet!

The most important overall success factor was the requirements engineering phase. Models and interviews were used to gather requirements from a variety of stakeholders. The requirements were then presented in 3 complementary views.

As in the requirements phase, the key to successful specification was to use the right combination of methods. User interfaces were specified by iterative prototyping with close end-user involvement. The data handled by the system was precisely described using VDM. Formal proof was reserved for the most critical concurrent elements.

The formal specification was refined until the final mapping to C was relatively straightforward and code could be produced using standard templates. This mitigated one of the risks of using C: that programs are poorly structured and hence that it is difficult to show compliance with the specification.

The results of our approach speak for themselves. Independent assessment (Pfleeger & Hatton 1997) has shown that the quality of the code is far better than the industry average for new developments. In the nine years since its commissioning, CDIS has since shown a very high level of operational availability. The robustness of CDIS and its conformance to its specification stands in great contrast to that of the other systems developed for LATCC.

5.2 Example: Zero Defects Post-Delivery

The major difference between a thing that might go wrong and a thing that cannot possibly go wrong is that when a thing that cannot possibly go wrong goes wrong, it usually turns out to be impossible to get at and repair.
Douglas Adams, "Mostly Harmless", October 1993

In the mid-1990s Praxis was subcontracted by Power Magnetics and Electronics Systems (now part of Ultra Electronics) to develop the application software for the UK MOD's Ship Helicopter Operating Limits Information System (SHOLIS). This is a SIL-4 system which aids the safe operation of helicopters on naval vessels (King et al. 2000). The main features of SHOLIS are:

- First software to UK MoD Interim Defence Standard 00-55
- SIL-4 embedded information system
- Small team
- Formal specification in Z
- Proof of specification most efficient means of eliminating faults
- Written in SPARK Ada (27,000 lines)
- Proof of exception freedom for most of the code
- 0 defects in sea trials

Defence Standard 00-55 (MOD 1991) is among the most stringent procurement standards in the world. It requires a formal specification and design, with formal arguments to link the specification, design and code. Many suppliers claimed that the level of formality required was unrealistic. Praxis was the first company to show that it is not only possible but desirable to construct software to such demanding standards.

Our approach included:

- A small team using maximal tool support
- The most powerful workstations available, because "a big computer is far cheaper than the time of the engineers using it"
- A simple system architecture which made the Z proof feasible
- Some 150 Z proofs carried out covering 500 pages, finding 16% of the faults identified pre-delivery for only 2.5% of the overall effort

- Selective partial code proof strategy, with additional manual analysis to show termination of all loops, functional separation and correctness of SIL-4 components
- 9000 verification conditions proven, most of them automatically

Most significantly, we found that Z proof was one of the cheaper stages at which to fix faults because it was early in the life-cycle, before any significant design activity had been undertaken. The next most efficient phase was the system validation test, which took place after integration, and was therefore one of the most expensive stages at which to fix faults.

Despite the unusually large amount of proof activity on this project, our approach was pragmatic, limiting the proofs to only those areas which were required to demonstrate compliance with Defence Standard 00-55.

5.3 Example: Halving The Cost of V&V

Praxis provided the majority of the software design team for a stores control unit for use on Royal Navy submarines. This is part of the UK's Submarine Acoustic Warfare Command System (SAWCS), which is to provide the Royal Navy with an effective defence against the latest generation of torpedoes. The main characteristics of this project are:

- Defence Standard 00-55 SIL-3 development
- Small team
- Innovative technical plan
- Use of INFORMED design methodology
- Half the normal effort for V&V

Traditional software developments view the testing phase as a means of finding faults. The problem is that testing comes too late: it can be orders of magnitude more expensive to fix faults late in the life-cycle (Boehm 1976, Baziuk 1995, Leffingwell 1997). Testing can thus be seen either as a very expensive way of eliminating faults, or as a relatively cheap way of confirming that the system has been built correctly.

Our experience on SHOLIS, where unit testing took 25% of the effort, but was one of the least cost-effective ways of finding faults, suggested that it might be better to avoid unit testing all together. *We were only able to do this because the software was specified in Z and implemented in SPARK Ada.*

We defined our development process to:

- produce software that would be easy to verify;
- produce certification evidence as a by-product;
- minimise reliance on unit testing;
- achieve structural test coverage by system testing, as recommended by DO-178B (RTCA-EUROCAE 1992); and
- take a system safety engineering approach, identifying the tension between maintenance and operational requirements.

The result was that the project expended 29% of total effort on V&V, compared with a norm of 40-60% for SIL-3 / SIL-4 systems.

5.4   Example: First Ever Certification to ITSEC-E6

Your approach to requirements [REVEAL] was very effective.
John Beric, Head of Security, Mondex International

I've never seen such a clean hand-over of deliverables.
Andy Calvert, Security Development Manager, Mondex International

Mondex International (MXI) invented the Mondex Purse, an electronic smartcard alternative to cash. Praxis worked with MXI to develop the MULTOS Global Key Centre (MGKC), which generates and manages the cryptographic keys used by the MULTOS operating system to maintain security.

MULTOS and the Mondex Purse were the first consumer products to be certified to the extremely demanding UK ITSEC-E6 standard. Three years on, only one other product has achieved this. The key features of MGKC are:

- Critical component of ITSEC-E6 system
- Capture user requirements
- Formally specified
- Mixed-language implementation - SPARK, Ada, C++, SQL - 100,000 lines total
- 0.04 defects per KLOC

The key points of our development process were:

- capturing the user requirements for the system using our requirements engineering method, REVEAL;
- use of our design method, INFORMED, for the security-critical parts of the software architecture and design;
- implementing the user interface in Visual C++, database queries in SQL and the critical components of the system in SPARK Ada; and
- controlling the risks introduced by a mixed development environment.

The vast majority of defects in the system were eliminated soon after introduction. Only 6% of the total project effort was spent fixing faults. Only 4 faults were found in the year post-delivery, a defect rate of 0.04 per KLOC (Hall & Chapman 2002) – lower than the Space Shuttle software (Joyce 1989), but much cheaper.

6   Conclusion

We have debunked the common myths surrounding the difficulty of producing high-quality software at a reasonable cost. Although current industry practice is to produce complex and expensive projects prone to failure, and the common perception is that this is inevitable, this perception is unjustified. The key to success lies in treating Software Engineering as a true engineering discipline.

The white box software development approach described here has been applied to a range of commercial projects with successful commercial and technical results. There is no "silver bullet" in this approach; it concentrates on identifying the right people, processes and tools for the job. We spend less time and money fixing faults as a result of getting the software right in the first place.

References

Amey, P. 2001. "Logic versus magic in critical systems." In *Lecture Notes in Computer Science* vol. 2043, pub. Springer-Verlag.

Baziuk, W. 1995. "BNR/NORTEL path to improve product quality, reliability and customer satisfaction." In *Proceedings of the 6th International Symposium on Software Reliability Engineering*.

Boehm, B. W. 1976. "Software Engineering." In *IEEE Transactions on Computing and Software Engineering*, 1:1226–1241.

Brooks, F. P. 1986. "No silver bullet – essence and accidents of software engineering." In *Information Processing* 86:1069–1076.

Brooks, Frederick P. 1995. *The mythical man-month: essays on software engineering.* Anniversary (2nd) edition, pub. Addison Wesley Longman, Inc.

Chulani, S. et al. 2001. "Deriving a Software Quality View from Customer Satisfaction and Service Data." In *Proceedings of ESCOM 2001.* Published on the internet, URL http://www.escom.co.uk/conference2001.

Committee of Public Accounts. 1993. "Sixty third report: Wessex Regional Health Authority Regional Information Systems Plan". The House of Commons.

Compaq. 1999. "Program license agreement for UK." Compaq Computer Corporation.

Hall, Anthony & Roderick Chapman. 2002. "Correctness by Construction: Developing a Commercial Secure System." In *IEEE Software* Jan/Feb 2002, pp. 18–25.

Hall, J. A. 1996. "Using formal methods to develop an ATC information system." In *IEEE Software* 130:66–76.

Hansard 2000. Parliamentary debate, Monday 24th January. *Hansard*, 6th series, vol. 343, 5th volume of session 1999-2000

Joyce, E. J. 1989. "Is error-free software achievable?" In *Datamation* 35(4):53–56.

King, S. et al. 2000. "Is proof more cost-effective than testing?" In *IEEE Transactions on Software Engineering*, 26:675–686.

Leffingwell, D. 1997. "Calculating your return on investment from more effective requirements management." Available from Rational, URL http://www.rational.com/media/whitepapers/roi1.pdf

Leveson, Nancy G. 1995. *Safeware: system safety and computers*. Addison-Wesley.

Lions, J. L. 1996. *Ariane 501 inquiry board report*. European Space Agency.

McConnell, S. 1999. *After the gold rush: essays on the profession of software engineering*. Microsoft Press.

MOD. 1991. "The procurement of safety critical software in defence equipment, INTERIM DEF STAN 00-55, Parts I and II". United Kingdom Ministry of Defence.

Ould, Martyn A. 1999. *Managing Software Quality and Business Risk*. John Wiley & Sons.

Pfleeger, S. L. & L. Hatton. 1997. "Investigating the influence of formal methods." In *IEEE Computer* 30:33–43.

RTCA-EUROCAE. 1992. "DO-178B / ED-12B Software Considerations in Airborne Systems and Equipment Certification." RTCA-EUROCAE.

Stålhane, T. et al. 1997. "In search of the customer's quality view." In *Journal of System Software* 38:85–93.

Standish. 1995. "The CHAOS report." The Standish Group, URL
`http://www.standishgroup.com/`

Standish. 1999. "CHAOS: a recipe for success." The Standish Group, URL
`http://www.standishgroup.com/`

Stephenson, A. G. et al. 1999. "Mars climate orbiter mishap investigation board phase 1 report." NASA.

Tassey, G. 2002. "The economic impacts of inadequate infrastructure for software testing." NIST. URL
`http://www.nist.gov/director/prog-ofc/report02-3.pdf`

Taylor, A. 2001. "IT projects sink or swim." In *BCS Review* pp. 61–64.

Reforming the Law on Involuntary Manslaughter: The Government's Proposals on Corporate Killing

Prashant Popat and Roger Eastman
Henderson Chambers
2 Harcourt Buildings
Temple, London EC4Y 9DB

What is manslaughter?

1. Manslaughter is a common law offence. It can be divided into 2 categories. "Voluntary" manslaughter is the description given to an offence where there is proof of intention to kill or cause serious injury and the offence would be one of murder but for mitigating circumstances such as provocation or diminished responsibility. This does not concern us today. "Involuntary" manslaughter occurs when someone kills another person but did not intend to cause death or serious injury but was blameworthy in some other way. This could arise where the accused was reckless as to whether another was killed or acted in a grossly negligent manner that led to the death. It is this form of manslaughter that the government is proposing to reform.

2. Corporations can be liable for involuntary manslaughter if an individual is identified as the embodiment of the company, its directing mind, and that individual himself is guilty of manslaughter. Where such an individual is convicted the company can be regarded as having acted recklessly or in a grossly negligent manner and to have caused the death. The requirement to identify such an individual is known as the doctrine of "identification".

Difficulties in proving corporate manslaughter

3.   The need to identify an individual in a corporation who can be described as a directing mind, who was grossly negligent and whose negligence actually caused the death in question can be very difficult. Due in no small part to the skill of company secretary's in setting up an appropriate corporate structure!!

4.   In large corporations the power and control usually rests with board directors or very senior managers. They are unlikely to have sufficient involvement in the day to day activities of the corporation to enable the prosecution to show they were responsible for the relevant management or system failure. The more diffuse the corporation structure, the wider the company's activities and the larger organisation the harder it is to mount a successful prosecution.

5.   The consequence of these difficulties is that there have been few prosecutions for corporate manslaughter and only 4 successful prosecutions. Each of these involved small companies where the directing mind had substantial involvement in the actual business of the corporation.

6.   A decision in the last year lent some support to the argument that the identification principle does not apply in every case when, a small company, English Brothers Limited, (EBL) pleaded guilty to an offence of manslaughter notwithstanding that no individual directing mind was convicted of the offence. However, in that case the person who was probably the directing mind died before the prosecution was commenced and the guilty plea may have been entered on the basis that an individual had been identified, albeit he could not be prosecuted.

7.   The traditional view was re-affirmed by the events in the "Simon Jones" case. This case concerned the death of a young man, Simon Jones, who was sent by an employment agency to work for Euromin Ltd who undertook dock work. Mr Jones was required to take part in the unloading of cobblestones from the hold of a ship. The operation involved the use of a crane supplied with both a grab bucket for lifting loose material and a lifting hook for lifting goods contained in bags. On his first morning, Simon Jones was working beneath the grab bucket. He had received no training and was not warned of the potential hazard of working beneath the grab bucket. In the driver's cab, the operator inadvertently pushed the lever the grab bucket closed and immediately decapitated Mr Jones.

8.   Following a lengthy police investigation, the CPS announced that it would bring no charges. Following a challenge from Mr Jones' family, a new

caseworker reached the same decision. On this occasion Mr Jones' family sought judicial review before the High Court on the basis that the CPS had applied the wrong test for manslaughter when reaching their decision. The High Court agreed finding that the CPS erred in applying a test of subjective culpability rather than an objective test required following recent decisions. They confirmed the need to identify the directing mind.

9. The objective test appears to be the test used before the Central Criminal Court when the trial was heard late last year. The jury found the Managing Director not guilty of manslaughter and consequently Euromin were also acquitted.

Demands for change

10. The number of higher profile disasters where there was no prosecution for corporate manslaughter or no successful prosecution has led to much public dissatisfaction. The Herald of Free Enterprise, the Kings Cross Fire and Clapham are but 3 examples.

11. These difficulties led to the Law Commission's Proposals in their report no. 237 *"Legislating the code: involuntary manslaughter"*. These proposals recommended a new offence of corporate killing which focussed, essentially, on the corporation as a whole. So that if it could be said that the corporation was (grossly) negligent in the way it might be in a civil claim, the corporation could be liable for the offence of corporate killing. The essential requirement of the proposed offence was that there had been a management failure which caused death.

12. The Commission's proposals received much publicity and appeared to have political support but they were not actioned and the common law offence continues to apply.

13. Following the Southall crash in 1997 the train operator, GWT, was prosecuted for corporate manslaughter. The prosecution sought to argue that the elements of the offence of corporate manslaughter had changed following a decision of the House of Lords[1] in the prosecution of an individual for manslaughter. It was argued that there was no longer any need to identify a controlling mind who was also guilty of manslaughter. It was possible, the prosecution contended, to convict the company if there had been a serious

---

[1] R v Adomako [1995] 1 AC 17

breach of its personal duty to the deceased. The prosecution failed at first instance and in the Court of Appeal. The court of Appeal re-affirmed the identification principle and the need to establish the guilt of a human individual before a non-human could be convicted.

14. The failure of the GWT prosecution and the subsequent rail disasters at Ladbroke Grove, Hatfield and Potters Bar have intensified the demands of victims for a change in the law of corporate killing. It is believed by many that there is a pressing need to hold corporations criminally accountable for the consequences of "accidents" and some believe that the threat of such accountability will lead to improved safety. It is these demands and these beliefs that have led to the Home Office proposals published in May 2000 and entitled: *"Reforming the law on involuntary manslaughter"*.

15. The public clamour for such change has continued unabated since the publication of the Home Office proposals. The demand seems to be shared by Director of Public Prosecutions, the HSE and the TUC all of whom have commented on the inadequacy of the law following the failure of the prosecution arising out of the death of Simon Jones. Simon Jones' family has set up a pressure group to press for legislative change. Their case was recently taken up by the Mark Thomas Project, a Channel 4 television programme.

The Home Office proposals

16. The Home Office proposals on voluntary homicide propose new manslaughter offences and other associated offences. The proposals build upon the report of the Law Commission. If implemented they will no doubt meet most of the public's demands for retribution; whether these proposals achieve what I suggest should be their primary aim viz, the improvement of safety, is another matter and one which I shall consider further when considering the major proposals.

17. These proposals affect the offences of involuntary manslaughter against the individual as well as corporations. For the purposes of this discussion, however, I will focus on the proposed new offence of corporate killing and the proposals associated with that offence.

Proposed offence of corporate killing

18. By the draft bill[2] a corporation will be guilty of corporate killing if:

    (a)   A management failure by the corporation is the cause or one of the causes of a person's death; and

    (b)   That failure constitutes conduct falling far below what can reasonably be expected of the corporation in the circumstances.

19. There will be a "management failure" by a corporation if the way in which its activities are managed or organised fails to ensure the health and safety of persons employed in or affected by those activities and such a failure may be regarded as a cause of a person's death notwithstanding that the immediate cause was the act or omission of an individual.

20. The effect of the proposal is to render companies liable to conviction if they fail to have a proper system in place and that failure leads, ultimately, to the death of an individual.

21. It will remain necessary for the prosecution to show, effectively, gross negligence on the part of the corporation because what is required is that the corporation's conduct "fell far below what could reasonably be expected".

22. The removal of the identification doctrine will materially increase the potential for corporate prosecutions and increase the potential for successful corporate prosecutions. Every case in which a civil claim is made seeking damages for negligence is potentially a prosecution for corporate killing if a fatality has occurred and the negligence could be described a serious. The same is true vice versa.

23. However, unlike the proposed offence of killing by gross carelessness, corporate killing does not require that the risk be obvious or that the defendant was capable of appreciating the risk. A grossly negligent management failure that leads to the death of an individual will be sufficient to convict the corporation even if the risk was not obvious. It is difficult to see how an organisation that has otherwise conducted an appropriate risk assessment of the risks created by its undertaking and implemented an appropriate control measure could be prosecuted in respect of a risk that was not obvious. However, the obviousness of the risk has been expressly excluded from the requirements of the offence.

---

[2] Clause 4

24. A further point made against the proposal is that it does no more than replicate the offences under s2 and 3 of the HWSA 1974 with the same penalty; an unlimited fine.

Remedial orders

25. In addition to an unlimited fine which is the available punishment for the offence of corporate killing, the government proposes that a court may make remedial orders against the corporate defendant. Such orders would be akin to enforcement and prohibition notices served by HSE under the existing health & safety legislation. Breach of any such remedial order would in itself be an offence and be punishable by a fine.

26. This proposal gives rise to concern. In relation to most accidents there is a need to ensure that any remedial orders are consistent with the strategy for improving safety across the industry. There would be a danger that a criminal court considering one aspect of the management of safety by one undertaking in that particular industry would be unaware of the impact of any remedial order on the activities of other industry operators and/or would be unaware of any advancements in safety being made elsewhere.

27. By way of hypothetical example, assume that a drug manufactured by a respectable pharmaceutical company caused the death of an individual. Assume further that the manufacture of the particular batch that contained the offending drug was grossly negligent. It is not fanciful then to assume that a criminal court that has convicted that pharmaceutical company of corporate killing might make a remedial order regarding the manufacture of the drug. Such a remedial order may in fact lead to a restriction or reduction in the production of the drug and that may cause harm to a large number of people. The merits of a remedial order in such circumstance would necessitate a complicated risk benefit analysis which a criminal court may be ill-equipped to conduct.

28. Further if the court trying the criminal proceedings were to be made aware of the "wider picture" in all cases where a remedial order was proposed, there is a potential for serious delays to the conclusion of those proceedings and an escalation (and possibly a duplication) of costs.

Disqualification order

29. The government has also raised or proposed other possibilities for enforcement action against a director or other company officer or employee. The first proposal is that any individual who can be shown to have had some influence on the circumstances in which a management failure, falling far below what could reasonably be expected, was a cause of a person's death, should be subject to a disqualification from acting in a "management role in any undertaking carrying on a business or activity in Great Britain".

30. This is a very wide-ranging disqualification order which would preclude any director (and possibly other employees) from having any managerial role, not just as a director of a company, in any subsequent business. It would also not be necessary for the prosecuting authority to show that such a director was solely responsible or even primarily responsible for the management failure. The disqualification order can be for a limited or indefinite period of time.

31. Again it is submitted that this proposal goes too far and is unnecessary. Any conviction of a corporation for the offence of corporate killing, particularly a large company, will lead to serious repercussions and consequences for that corporation and the relevant managers in any event. The consequential effect on the value and financial performance of that organisation and the prospects of those managers is in itself a sufficient deterrent.

32. Insofar as it is necessary to protect the public against subsequent acts of individuals whose activities are covered by the provisions of the companies acts, the courts already have a power to make an appropriate disqualification order.

33. Further, any application for a disqualification order could be made only after the conviction of the corporate defendant. The trial of any such application would have to be preceded by the giving of particulars and the provision of relevant evidence. The usual procedure would therefore be that such proceedings would be brought some time after the conviction of the corporate defendant. History suggests that the conclusion of any criminal trial for the offence of corporate killing will take place months, if not years, after the accident in question (the Southall trial, for example, took place almost two years after the crash). There would then be a further delay before proceedings were instituted against the directors or company officers and a further period of months if not years before those proceedings came to trial. There would therefore be a delay in the conclusion of proceedings arising from any incident and directors and officers would have the threat of any

such proceedings hanging over their heads for some considerable time, likely amounting to years.

34.   If it were proposed to bring disqualification proceedings separate from, or in the absence of, any prosecution for corporate killing there would be substantial difficulty for the parties and the courts. The contribution of the director or other company officer can only be determined and considered in light of the corporation's alleged failing and in light of the causative link between that failing and the death of the individual. This would effectively mean that there would have to be a trial in the absence of a corporate defendant to establish its guilt for the offences of corporate killing. This would be unfair to the directors or officers and would lead to complicated and lengthy trials.

35.   The need to protect the public against the actions of an individual, and the need to deter individuals from mismanagement which leads to death, are adequately protected by the proposals for the individual manslaughter offences and the availability of disqualification proceedings under the companies act.

36.   Further these proposals would probably preclude many individuals who have been subjected to a disqualification order from ever working again. The individuals likely to be covered by such an order will be persons who have held managerial posts (perhaps senior posts,) in the corporate undertaking. The effect of the order would probably be to deny them any opportunity of working in a similar post in that or any other organisation and would, therefore, probably prevent them from finding alternative employment. In my view, the public could be protected by requiring the individual to demonstrate a safe and effective system of management e.g. By providing regular reports to the enforcing authority.

Offence of management failure

37.   A further proposal is that where an undertaking is guilty of corporate killing there should be an additional criminal offence so that officers of the undertaking who contributed substantially to the management failure resulting in death are liable to a penalty of imprisonment or a fine in separate criminal proceedings. Such a proposal would introduce a new offence and cover the situation where a director could not be convicted of reckless killing or killing by gross carelessness but nonetheless played a substantial role in

the management failure that led to the death of an individual and the undertaking itself was convicted.

38. The points made under the previous heading are repeated in respect of the proposal that officers of undertakings, if they contribute to the management failure resulting in death, should be liable to a conviction and a penalty of imprisonment.

39. Such an offence would be unfair to those persons involved in entrepreneurial activity. There is little justification in principle for singling them out and exposing them to the risk of a conviction of a criminal offence in respect of acts or omissions that will otherwise not warrant a criminal conviction. If their acts or commission are sufficiently serious and causative of the death they will undoubtedly be liable for one of the individual manslaughter offences; otherwise they should not be in a different position from every other member of society.

40. It should also be recalled that the enforcing authorities have powers under section 37 of the HAWSA 1974 which are widely drawn and which will cover the acts or omissions of officers of undertakings which have exposed individuals to risk of harm.

Discussion

41. The question of corporate accountability has been the subject of much publicity and debate in relation to accidental fatalities and this is understandable. The current proposal would lead to substantially greater prospects of criminal prosecutions of companies and their managing officers.

42. In my view, however, the current proposals go too far in some respects and may lead to the promotion of a culture which seeks to apportion blame and to hide mistakes. This would lead to activity designed to give the impression of good safety management by the creation of audit trails for the prime purpose of exculpation. The primary focus of the proposals for change to the law of manslaughter should be to improve safety. The principles of retribution and deterrence are secondary.

43. The effect of these proposals will be to create a range of penalties which will cause individuals to be unnecessarily circumspect in the conduct of their duties and that will cause them to deny mistakes or seek to transfer blame.

44.   The danger of over-emphasising the criminal consequences of mistakes was considered in a recent report published by the department of health entitled *"An organisation with a memory"*. The purpose of this report was to learn lessons from adverse events in the NHS. We believe that the authors of that report correctly identified the need to move away from a blame culture if appropriate lessons are to be learned and if safety is to become paramount. The following passage comes from the executive summary to that report under the heading of *"Evidence and Experience"*

> *"7.   When things go wrong, whether in health care or in another environment, the response has often been an attempt to identify an individual or individuals who must carry the blame. The focus of incident analysis has tended to be on the events immediately surrounding an adverse event, and in particular on the human acts or omissions preceding the event itself.*
>
> *It is of course right, in health care as in any other field, that individuals must sometimes be held to account for their actions - in particular if there is evidence of gross negligence or recklessness or of criminal behaviour. ...*
>
> *Activity to learn from and prevent failure therefore needs to address their wider causes. ...*
>
> *It is possible to identify a number of barriers than can prevent active learning from taking place, but there are two areas in particular where the NHS can draw valuable lessons from the experience of other sectors.*
>
> ***Organisational culture** is central to every stage of the learning process - from ensuring that incidents are identified and reported through to embedding the necessary changes deeply into practice. There is evidence that "safety cultures", where open reporting and balanced analysis are encouraged in principle and by example, can have a positive and quantifiable impact on the performance of organisations. "blame cultures" on the other hand can encourage people to cover up areas for fear of retribution and act against the identification of the true causes of failure, because they focus heavily on individual actions and largely ignore the role of underlying systems. ...*
>
> ***Reporting systems** are vital in providing a core of sound, representative information on which to base analysis and recommendations. Experience in other sectors demonstrates the value of systematic approaches to recording and reporting adverse events and the merits of quarrying information on "near misses" as well as events which actually result in harm..."*

45. It is submitted that the current proposals could affect organisational culture and reporting systems in exactly the way feared by the authors of this report.

46. However, whatever the merits or otherwise, of these proposals it is plain that companies will need to start putting in place appropriate safeguards, audit processes and protocols for the recording of all possibly material decisions. Changes to the law on corporate killing are very likely and it is further likely that these changes will make it easier to mount prosecutions. The current proposals are being reviewed in light of the consultation process. We understand that the Home Office received a substantial number of responses and that it is planning to publish a report on its web site in the next few weeks, if not days, setting out the tenor of the responses received and its current understanding of the date of implementation. The best guess at the moment is that it will be included in the Parliamentary session from November 2002.

THE SAFETY CASE – 2

Electronic Safety Cases:
Challenges and Opportunities

Trevor Cockram   Ben Lockwood
Praxis Critical Systems Limited   Raytheon Systems Limited

Abstract

This paper describes the use of electronic formats for safety cases to meet the requirements of a number of military and civil standards. The challenge to safety engineers is to produce safety cases that are quickly readable, intelligible and auditable even when a large amount of material is required. We describe the problems in developing complex safety cases using traditional development methods and the opportunities to address these problems by the development of an electronic safety case. We then describe an example eSafety Case and how this can be used to manage a safety programme and to produce a safety case that will meet the requirements of the certification authorities.

1 Introduction

The provision of a safety case is a requirement of many standards [HMSO 1992, MoD 1997, MoD 1996, HSE 2000]. A Safety Case presents the argument for the safety of a system and summarises and justifies the supporting evidence. A Safety Case is an input to the safety approval process for a system. The body responsible for safety approval will consider all relevant safety submissions, primarily the Safety Case and supporting documentation such as reports of Safety Audits and Assessments, in order to satisfy themselves that the system is adequately safe and conforms to the relevant international, national and industry safety standards.

The detailed content of a safety case can vary, but in this paper I am referring to the body of information that makes the case that the system is safe. This includes:

- defining the system, including the system boundary and the system architecture and functionality;
- identifying the development and safety management plans;
- providing the logical argument that shows that the system is safe based on the evidence available;
- detailing the hazard and accident identification, causal and consequence analysis;
- demonstrating that risk reduction has been carried out to an acceptable level, with resultant closure of hazards;
- defining safety requirements and methods of verification; and
- describing any limitations and caveats; and the final conclusion on the safety of the system.

The provision of safety cases has resulted in an ever increasing workload for equipment producers (to produce the safety cases and manage their contents), and also for independent assessors and regulators (who are required to assess safety cases). Typically a safety case for a moderately sized system, along with the reports that constitute the primary supporting evidence, can result in a pile of paper several inches thick (larger systems safety cases have filled library shelves).

The challenge to safety engineers is to produce safety cases that are quickly readable, understandable and auditable even when a large amount of material is required. Fortunately opportunities exist to de-risk the safety case development process by presenting the safety argument for review early in the programme and building up evidence incrementally throughout the development and use of the system.

2 Safety Case Problems

The problems of developing and using a safety case, apart from the obvious one of carrying out adequate safety analysis and hazard mitigation, include the following:

2.1 Paper-based Safety Cases

2.1.1 Ease of Navigation

A typical safety case for a single piece of equipment is of the order of six inches thick and starts with hundreds of pages of system description. This can have the effect of immediately switching off the reader, as it is not clear where the relevant safety information can be found.

2.1.2 Ease of Drilling Down

Another issue in paper safety cases is the difficulty of finding a trace through the information provided, for example the accountability and competency for the safety decisions made.

2.1.3 Configuration Management/Control

It is often the case that the material with the safety case comes from a number of different processes and tools. This can lead to a major document management task and difficulty in generating and maintaining a common baseline for the material. The issues include unique identification of, and access to, documents (including specific versions of them). There is also a potential problem with organising the information with sufficient indexing and referencing mechanisms to allow the reader to find all the relevant information.

2.1.4 Diversity of Information Packaging

It is also difficult (or even impossible) to maintain a consistency of presentation. A problem often comes with the use of a proprietary hazard log tool that is only capable of producing reports in a single format that is impossible to embed within another document. This often results in additional summary hazard reports being written to overcome these shortcoming, but result in an amount of nugatory work.

2.1.5 Ease of Packaging and Delivery

Large paper-based safety cases are unwieldy, which results in large volumes of paper being generated, collated and being sent to the interested parties. The resulting safety case is bulky, not in an easily readable form and difficult to modify.

2.2 Clarity of safety argument

With many safety cases there is a problem with the clarity of the safety argument and links to the rest of the content of the safety case. The safety argument is often buried in sections of the detailed text and not easily traceable to the supporting evidence. The individual threads of the safety argument may also be dispersed through the levels of documentation provided.

2.3 Multiple certification standards and processes

Many systems are developed with the intention of dual or multiple uses. The consequence is that a system is certified for one use and there is a large translation task to make the safety certification and information acceptable to another certification body. In practice, the safety data required for different certification bodies is similar, i.e. to show that the likelihood of hazards has been reduced to an acceptable level and that the processes used for development are appropriate. The differences in certification compliance therefore can be addressed through the development of a cross-reference index for compliance to each standard.

2.4 Dealing with multiple variants

Much of our work is associated with the development of safety cases for equipment that is used in the rail industry, civil and military aviation (both on the ground and airborne), and the finance sector. The equipment is often characterised by large numbers of variants and applications. The safety management for the project needs to take account of the high degree of re-use within the equipment and functional designs to provide a consistent and cost effective means of conducting the safety management and the delivery of safety cases.

3 Overview of the eSafety Case Approach

With a large amount of material to manage, the intuitive approach is to treat the safety case as an intranet and to manage and link the material together by means of hypertext. The idea of using hypertext to link information with a safety case is not new, see for example [Brown R 1998], however, the challenge is to address the issues of complex safety cases in an efficient manner to provide a safety case which is manageable, auditable and acceptable to the certification authorities.

The overall approach is based on the use of an electronic presentation of the safety case using standard PC browsers and plug-ins. This provides a readily available and common platform for presentation – removing the need for bespoke tools.

3.1 The single document approach

One of the problems with current safety cases is that the information is generated throughout the system development; it is often diverse and presented in various formats. In addition, standards require that various safety documents be produced at stages of the development without necessarily a view to the ultimate goal of an acceptable safety case. The approach we have used is to use an electronic safety case that provides a single electronic document that presents the safety information for the system at various points during the programme lifecycle. This performs the traditional function of a safety case, but also contains each of the documents required during the safety lifecycle. By using a single evolving document like this, the consistency of safety information is maintained with a view of reaching the ultimate goal of an acceptable safety case. One of the benefits of a safety case with electronic links to the key supporting information during the development process is that the readers can see how the safety case is developing and take steps to plug any gaps which become apparent. By making the intentions of the safety case clear early in the system development programme, the risk of substantial rework to obtain certification at the end of the process is considerably reduced.

In this way the eSafety Case can perform the functions of the following documents:

- Safety Case, including safety argument and supporting evidence
- Safety Plan
- Preliminary Hazard Analysis
- System Hazard Analysis
- Hazard Log Report
- Safety Requirements

3.2 The Electronic Safety Case

The key navigational elements of the electronic safety case are:

- The overall presentational structure (chapters/sections of the safety case) presented through a menu bar.
- The safety argument (presented in Goal Structured Notation), which provides a means of navigating from a specific safety argument goal to the assumptions, strategy/justification and the supporting evidence.
- The hazard log (included in the safety case structure) that provides a traceable route to the individual assessments, mitigation approaches, safety requirements and evidence related to the specific hazards.
- Other hyperlinks used as appropriate from the detailed pages.

Information is generated in one of the following ways:

- General Textual or Graphical Information: produced using standard office tools and incorporated into HTML pages.
- Database or other Tool Outputs: many tools allow reports to be generated in an HTML format and these can be easily incorporated or massaged by standard scripts into usable components of the safety case.
- Reference to PDF documents: most systems using browsers are set up with the Acrobat Reader plug-in which allows reports to be easily included in the safety case format.

A particular safety case is based on:

- Common Skeleton: which includes all pages that can be reused for all variants and/or views (much of the common skeleton can actually be common to different programmes in our experience).
- Variant Specific pages: which address issues, methods or items which are only applicable to a particular variant of the system .
- View Specific pages: which address presentational issues with respect to a given safety case view (reflecting for example Def Stan 00-56 [Mod 1996] format, or a JSP 430 safety case format [MoD 1996-2]).

Clearly there is benefit in maintaining a significant amount of re-use through the common skeleton if variants or views are required. The development of our safety case is based on an incremental approach where:

- A skeletal safety case framework is built early on (partly off-the-shelf).
- A hazard log database is developed and maintained.
- An initial safety argument is developed.
- 'Builds' of the safety case are carried out at appropriate times during the programme to reflect the current snapshot of the safety case.

3.3 Improving the safety argument

Electronic Safety Cases provide improved safety arguments in several ways. The argument can be represented graphically in the form of a goal structured notation argument, which can be structured into a number of linked hypertext pages so that each section of the argument can be clearly seen. Supporting text can be provided on each page to provide a commentary on the graphical argument. Safety case navigation can also be improved by making the graphics hypertext sensitive so that the user can link directly to the evidence and data that support the argument.

3.4 Dealing with multiple certification standards

It is possible to generate alternative views in the format of hypertext navigational structures of the safety data held within the safety case to show the information in a form that the certification body finds acceptable. As an example, a system containing software has been developed to Defence Standard 00-55 [MoD 1995] SIL-2 requirements. One view of the safety case provides all the requirements of a safety case to this standard. Another view of the safety case shows a safety case for certification to RTCA DO178B level C [RTCA 1992]. The mandatory documents for civil certification, i.e. the Plan for Software Aspects of Certification, the Software Accomplishment Summary and the Software Configuration Index, uses the same information as in the Def-Stan 00-55 safety case but rearranged into a form to meet the civil certifiers expectation solely by means of hypertext links. Both the civil and military certification authorities have accepted this approach and noted that the electronic safety case made their jobs easier.

3.5 Dealing with multiple variants and applications

In projects where there are multiple variants and applications it is desirable that a common safety case be developed and modified as required. This implies that safety cases that are based on a re-use model would be the most efficient means of forming the large number of safety cases required for the equipment in the various variants and applications. We have developed a re-use model using the principles of domain analysis applied to the equipment requirements.

This analysis identifies the domains of the requirements, ie whether they are:

a) Common to all applications;
b) Specific to equipment variants and options;
c) Common to a generic application types;
d) Specific to a single application incidence.

Hypertext technology is used to manage the safety case arguments and supporting data in a configurable form. This approach allows a user of the safety case to browse it for the particular configuration of the equipment that is of interest. The configurable version of the safety case proposed is to be delivered either as a computer file which can be browsed by standard browsing tools such as Internet Explorer, or a printed copy of the output of the specific configuration of the equipment.

The contribution to the platform safety cases will follow a similar approach, i.e. taking a subset of the equipment safety case together with the generic platform and specific platform configuration

4. An example eSafety Case

One of the points about the electronic safety case is that it should be accessed via a computer and therefore in the printed form it is difficult to show its functionality. An example electronic safety case will be demonstrated at the Symposium. More on-line information about the example electronic safety case can be found at www.esafetycase.com.

4.1 System description

This is an eSafety Case produced for a totally fictitious system: The Totally Imaginary System (TIS) Programme is for the design, production and installation into system platforms of equipment that includes a secure encrypted radio, and to provide in-service support of the TIS equipment. The new equipment, consisting of Communicator and Transmitter products, will replace the existing equipment where fitted. The systems will include Frequency S functionality (selective, data-link mode) and have been designed with the capability to incorporate future technology.

4.2 Management summary

The first page of the electronic safety case shown in Figure 1 is the management summary that gives the background of the system, the standards to which the safety case is being developed, links to system description and argument, the conclusions and any limitation and caveats that apply to the safety case. You will also note in the layout that there is a navigation frame that appears in each safety case page to allow the user direct access to the principal sections of the safety case.

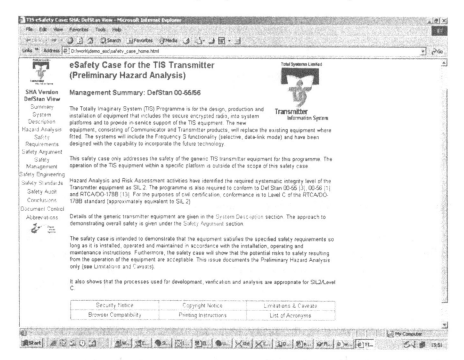

Figure 1. eSafety Case for the TIS Transmitter – Management Summary.

4.3 GSN fragments

The safety argument consists of two components within each page of the argument, a fragment of the goal structured notation diagram and the supporting commentary. Navigation through the argument is facilitated by active links within the GSN diagram and via the text. In practice a page of the safety argument will consist of a goal, its strategy, both supported by context statements and the supporting sub-goals. It is possible to provide hypertext links to supporting information in both the diagram and in the supporting text. For example in Figure 2 boxes in the GSN would be linked to more detailed supporting text or lower level diagrams such as the one shown in Figure 3.

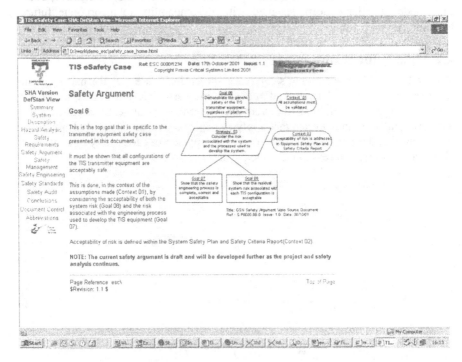

Figure 2. eSafety Case for the TIS Transmitter – Safety Argument.

At an early stage of the development, the safety argument will only be an outline of the final safety argument. It is useful to include potential solutions (or markers) for goals even though these solutions are at a high level. The review of the early safety argument can assist in both providing a sound framework for ongoing development and assist with 'buy-in' from the client and/or regulatory authorities.

We suggest that developing the safety argument as fully as possible at an early stage in the system development is a valuable way of managing the safety programme. It will identify what needs to be done to complete the safety argument and back it up with evidence. This helps to de-risk the safety case development programme.

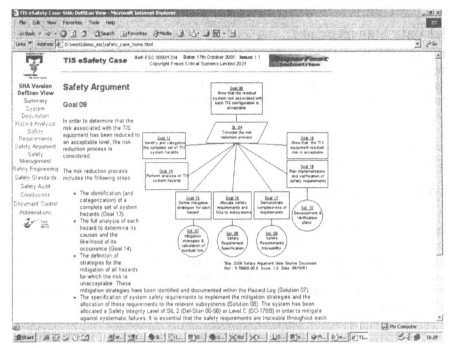

Figure 3. eSafety Case for the TIS Transmitter – Safety Argument.

5 Further considerations

Using an eSafety Case is not a silver bullet to solve safety issues. The following issues need to be considered before using the eSafety Case approach.

1. Only as good as the underlying SMS process.
2. Configuration and quality of data.
3. Review and user testing.
4. Tools and Build Automation.

5.1 The safety management process

An eSafety Case helps with the presentation of the safety argument and supporting information. It does not provide the means of generating the contents of the safety case. Applying an adequate safety management process is necessary to develop any safety case. The eSafety Case can have the effect of quickly and clearly identifying any weakness in the hazard identification process, the safety analysis and in the material that supports the safety argument.

5.2 Data Management

The material that makes up the eSafety Case requires careful configuration management to ensure that the links within the case point to the correct version of the data. Data within a developing project will change and the eSafety Case must reflect this. A suitable configuration management tool is essential to hold and control the data.

As data is used its quality must be assessed to ensure that the safety argument remains valid and correct.

5.3 Review and User Testing

Before an eSafety Case is released the safety case developer must comprehensively and systematically review it. The nature of an eSafety Case makes this difficult, however it is easy for the eSafety Case user to "drill" down through the links to follow a specific trail. A systematic process is therefore required to review the content of every page contained within the eSafety Case and testing needs to be undertaken to confirm that the case is readable and that the links are made to the correct place.

5.4 Tools and Build Automation

One of the philosophies behind the eSafety Case is to reduce the number of tools required to use the safety case to a minimum set which will normally be readily available on all PCs (for example Adobe Acrobat Reader and HTML browsers).

The developers may have preferences for certain tools (e.g. database and analysis tools). We have tried to separate out the management of items such as a hazard log database from the safety case build process by utilising HTML export and other filters to ensure that the main build and test process is not constrained by the underlying toolset.

We have extended this idea as far as possible in the generation of the eSafety Case, however, it is necessary to link the data from a number of different tools together. We also know that users will have preferences for particular proprietary databases, analysis tools etc.. The approach we have used is to automate the eSafety Case build process and to make it possible to tailor this build process as much as possible.

6. Conclusions

We have used this approach for the development of safety cases on a number of projects most notably the Successor Identification Friend or Foe programme where eSafety Cases have been produced and accepted by both the military and civil certification authorities against a number of different standards including Defence Standards 00-55 and 00-56, JSP430 and RTCA DO178B.

The response of both the certification authorities and system development teams to the eSafetyCase has been very positive. They have welcomed both the physical ease of use and the content presentation as a major advance.

The initial experiences using this approach have been encouraging. We expect to develop our supporting tools further to streamline build and test processes. We expect to continue learning from the feedback received on the current safety cases from both our immediate clients (developers) and their customers. In particular, we have received feedback suggesting:

- A rolling safety case (living as a document) could support the programme safety management activity as a planning and monitoring tool.
- Much of the structure and content of the safety case is generic and can be re-used for new programmes
- A similar approach could be adopted for other system documentation.

Acknowledgements

The authors would like to thank Ross Wintle for writing the software to allow many of these ideas to be implemented and John Harvey for helpful comments on the process and content of the paper.

References:

[Brown R 1998] - Improving the production and presentation of safety cases through the use of Intranet Technology R Brown in Industrial Perspectives of Safety-critical Systems ed Redmill and Anderson Springer 1998

[HMSO 1992] Offshore Installations (Safety Case) Regulations 1992.

[HSE 2000] Railways (Safety Case) Regulations 2000 Health and Safety Executive.

[MoD 1997] Ministry of Defence Directorate of Standardization. Defence Standard 00-55 Issue 2: The procurement of safety critical software in defence systems.

[MoD1996] Ministry of Defence Directorate of Standardization Defence Standard 00-56 Issue 2: Safety Management Requirements for Defence Systems.

[MoD 1996-2] Ministry of Defence Ship Safety Management Office JSP 430 Ship Safety Management System Handbook Volume 1 Issue 1

[RTCA 1992] Software Considerations in Airborne Systems and Equipment Certification RTCA DO178B Requirements and Technical Concepts for Aviation.

Safety Case Categories - Which One When?

Odd Nordland
SINTEF Telecom and Informatics
Trondheim, Norway

Abstract

The CENELEC railway application standards identify three categories of safety case, but give little guidance on which category should be used when. A brief description of the concept of safety cases is given, and the categories that the standard identifies are explained.

1. Introduction

Modern railway networks make extensive use of computer systems to control and monitor rail traffic in a safe and reliable way. A central element of such systems is the set of interlocking computers that control traffic and track-side equipment at stations and junctions. Here the term interlocking refers to the fact that any one computer must set the signals, points etc. that it controls in dependency of the settings at neighbouring locations. Each computer must therefore receive information from the neighbouring computers about the status of e.g. the signals and point switches that those computers control. In addition, it must get information about approaching and departing trains, and process this information together with its local data in order to set its own signals, point switches etc., and transmit appropriate data to the neighbouring systems for processing there. In addition, trains travelling in the area controlled by an individual interlocking computer must also receive up to date information.

On such networks, trains are also equipped with on-board computer systems that send data to and receive and process data from the interlocking computers. The kind of information sent to the train includes data about the train's maximum permissible speed, the distance to the next signal, the status of that signal etc. The on-board computer transmits data back to the interlocking computers, such as the train's identification, position, momentary speed and travelling direction etc.

Finally, there are rail traffic control centres that communicate with the interlocking computers to monitor and control traffic throughout the railway network, or at least a significant portion of it. If necessary, in suitably equipped networks traffic control centres can also intervene and stop trains directly. Thus, safe rail traffic is achieved with a highly distributed set of computers, both static and mobile, performing dedicated portions of the overall control task.

Clearly, the safety of a modern railway network is then highly dependent on the reliability and safety of all these computer systems and their interactions. Therefore it is natural to subject such systems to a thorough safety assessment before authorising their use. This together with European integration has given rise to a need for common principles and procedures.

2. Standards

One step in this direction is the adoption of European standards for railway applications, notably EN 50126 (CENELEC 1999), EN 50128 (CENELEC 2001) and prEN 50129 (CENELEC 2000). They describe processes to be followed in order to be able to assure the safety of a railway application. However, whilst they describe reasonably completely what to do, they do not go into great detail on how to do it.

Now of the above-mentioned standards, EN 50126 is the top-level document that covers the overall process for the total railway system. It defines Safety Integrity Levels and sets the frame for the more detailed activities that are described in EN 50128 and prEN 50129. prEN 50129 is the standard that defines the activities for developers and manufacturers, but also describes the requirements that a third party assessor shall verify. EN 50128 is the software specific "subset" of prEN 50129.

prEN 50129 requires that a safety case shall be submitted by a manufacturer and assessed by an independent third party before the safety authorities should approve commissioning the system. The term "Safety Case" is perhaps reasonably straight forward for people with English as their mother tongue, but experience shows a large degree of confusion when non English speaking Europeans use the expression! So the term requires some elaboration.

3. Safety Cases

The word "case" is used in a variety of contexts. We have special cases, briefcases, suitcases, court cases and - safety cases. The latter is derived from the concept of a court case: the prosecutor and defendant both "present their cases" to the court in order to convince the judge that their proposition is right.

Now for our purposes, the proposition is that a new (or modified) system is safe enough to use, and somebody has to present the case so that the "judge" - the safety authority - can reach a decision. As in legal proceedings, the "judge" will refer to an expert assessment by an independent party before relying on his own personal impression. (The word "assessor" did, in fact, originally mean "co-judge"!)

It is important to keep this analogy in mind. The safety case is a line of argumentation, not just a collection of facts. It aims at convincing a licensing authority, who does not necessarily have a deep knowledge of the specific technological questions involved, that a given product or system is safe enough to be taken into use. Or was safe enough: the safety case could well be used as evidence in genuine legal proceedings years after the system was authorised and commissioned.

The safety case must in itself contain enough information to give a clear impression of the system's safety properties and indicate where the details can be found if this is desired.

3.1   The Contents of a Safety Case

prEN 50129 gives a fairly extensive description of how safety cases should be structured, but the information is distributed through the standard and its annexes. Experience shows that there is a need for a briefer guideline.

In the following, the sections are briefly described. It should be noted that it is perfectly legitimate to let each section be a free-standing document, provided it has the right contents.

3.1.1 Definition of System

The first section in the safety case is the Definition of System. It shall give a complete and detailed description of the system for which the safety case is being presented. prEN 50129 states:
> *"This shall precisely define or reference the system/subsystem/equipment to which the Safety Case refers, including version numbers and modification status of all requirements, design and application documentation."*

It should be noted that this section is a definition of the system, not just a description. Its purpose is to identify exactly which version of the system is meant. Any modification of a component or module will necessitate a new - or at least updated - safety case. The same applies if the documents get updated.

3.1.2 Quality Management Report

This section is a report that describes what has been done to ensure that the system has the required quality throughout the entire life cycle. This involves:
- A description of the quality requirements with reference to the corresponding "source" documents.
- A description of the quality management system with references to the corresponding plans and procedures. In other words, a description of what one intended to do in order to ensure the necessary quality.

- A description of what actually was done, with references to e.g. audit reports, minutes of meetings and any other documents concerning the performed activities. In addition, any deviations from the plans and procedures shall be described and justified.

3.1.3 Safety Management Report

This is the corresponding report for safety management. As with the quality management report, the safety management report involves:
- A brief summary of the safety requirements with reference to the safety requirements' specification.
- A description of the safety management system with references to the corresponding plans and procedures. In other words, a description of what one intended to do in order to ensure safety.
- A description of what actually was done, with references to e.g. the hazard log, safety audit reports, test reports, analyses and any other documents containing evidence of the activities that were performed. In particular, the way the hazard log is handled must be described. In addition, any deviations from the plans and procedures shall be described and justified.

3.1.4 Technical Safety Report

This is the section where the technical safety characteristics of the system are described. It shall describe the underlying philosophy for achieving safety and identify which safety standards and design principles have been applied. The safety relevant properties of the system shall be presented and reference made to the corresponding evidence, i.e. test and analysis results, verification and validation reports, certificates and so on.

3.1.5 Related Safety Cases

If the system's safety relies on the use of safe parts or components, the corresponding safety cases shall be identified here. In such cases, any restrictions or application conditions mentioned in those safety cases shall be recapitulated here.

3.1.6 The Conclusion

This is more than just the statement that the product is sufficiently safe. prEN 50129 states:
> "This shall summarise the evidence presented in the previous parts of the Safety Case, and argue that the relevant system/subsystem/equipment is adequately safe, subject to compliance with the specified application conditions."

3.2   Different categories of safety cases

prEN 50129 identifies three different categories for Safety Cases, viz.:

- *Generic   product   Safety   Case   (independent   of   application)*
  A generic product can be re-used for different independent applications.
- *Generic   application   Safety   Case   (for   a   class   of   application)*
  A generic application can be re-used for a class/type of application with common functions.
- *Specific   application   Safety   Case   (for   a   specific   application)*
  A specific application is used for only one particular installation.

The underlying idea is actually fairly simple. It is assumed that railway applications are complex systems that consist of a number of less complex sub-systems that are configured to interact in a way that will fulfil the specified requirements. The sub-systems consist of even simpler "sub-sub-systems" that are configured to interact appropriately, and so the structure goes down until we reach a bottom level with "simple" products that cannot sensibly be subdivided further (i.e. we're not necessarily down to nuts and bolts). We start by "proving" that the simple products are safe, regardless of the configurations in which we use them. We then show that the configurations will be safe, provided we use safe products or configurations of products. And finally, we show that a specific configuration of explicitly identified products (or configurations of products) must be safe, because the (configurations of) products that it uses are safe, the underlying (system) configuration is safe and the way the products were incorporated into the system is safe.

In this context, the Generic Product Safety Case (GPSC) will present the safety case for a product, regardless of how it is used, so that it can be deployed in a variety of different safety related applications. The Generic Application Safety Case (GASC) will present the safety case for an application (configuration of products) without specifying the actual products to be used as components. It will simply refer to the generic properties that the products should have.

Finally, the Specific Application Safety Case (SASC) presents the safety properties of a particular combination of products in a given application. It will, of course, draw on the underlying GPSCs and GASC, but in particular, the details of planning and operation will be relevant here. In fact, prEN 50129 prescribes *"separate Safety approval ... for the application design of the system and for its physical implementation... "*, so there must be two SASCs:

- A "SASC - Design" that presents *"the safety evidence for the theoretical design of the specific application"*.
- A "SASC - Physical implementation" that presents the safety evidence for *"e.g. manufacture, installation, test, and facilities for operation and maintenance"*.

Unfortunately, the boundary between GPSC and GASC is rather fuzzy, because, as was pointed out above, a complex system that can use generic products (and therefore has a GASC) can itself be deployed as a " product" in a greater, more complex system. Then the "generic application" becomes a "generic product" within the more complex application. And then a specific application that is based on that generic application becomes - what?

At this stage it becomes evident that the expressions "generic product", "generic application" and "specific application" are highly context dependent. For the purposes of producing an appropriate safety case, the name should not play a significant role, but the contents of the safety case will vary according to the context (or category, to use the term in the standards).

4. Generic Product Safety Case

The expression "Generic Product Safety Case", that is used in the standard, is in itself somewhat misleading, because the word "generic" actually refers to the safety case, not to the product! In spite of the fact that the standard mentions "generic product", it is in fact a "Generic - Product-Safety-Case" and not a "Generic-Product - Safety-Case". Put another way, we have a "generic safety case for a product", not a "safety case for a generic product".

That being said, we still have a problem. Safety depends on the context in which a product is used, so presenting a generic safety case for a product without taking into account the way it will be used, i.e. its application, would appear to be a somewhat pointless task.

Take for example a microprocessor. What can be "dangerous" about a microprocessor, when you try to ignore the way it will be used? Well, it certainly has fairly sharp wires sticking out, so it might prick your fingers if you're not careful, and it could certainly do some harm to you if you swallow one. But does that require a safety case? Of course not.

However, if you intend to use a microprocessor in a safety related application, you will be able to say something about how suitable the microprocessor will be without having to go into details about the actual application. You could, for example, say something about the microprocessor's ability to perform calculations at a speed that will normally be more than fast enough for any safety related application, its reliability in performing those calculations accurately, and how robustly it will react to adverse conditions. That's what the generic safety case will be about!

In our example, you will also identify limitations to the type of safety related application that the microprocessor may be used in. This could for example relate to the processor's robustness, the need for a stable power supply, the necessary reliability of its periphery or even the maximum calculation rate for which accuracy can be guaranteed. These limitations will be demands on the application, i.e. conditions that must be fulfilled in <u>any</u> safety related application if the microprocessor is to be a genuine contribution to that application's safety. This is what the standard calls "*safety related application conditions*"!

You can also say something about the quality assurance routines that were applied in order to ensure that each microprocessor that leaves the factory has the specified properties. You can say something about the design principles that were applied and why they guarantee that the microprocessor will be the excellent product you claim it to be, and why this makes it suitable for use in safety related applications. And if you had potential safety related applications in mind when you designed the microprocessor, you can also say something about what you did to achieve those safety relevant properties.

So even without knowing anything about the intended application, you have all you need for a full scale generic safety case for your microprocessor. Or any other not too complex product: when things get complex you'll probably end up with a generic application.

5. Generic Application Safety Case

In the previous section it was pointed out that "generic product safety case" should really be "generic safety case for a product". For a "generic application safety case" the situation is the opposite: this is truly a safety case for a generic application. So what is a generic application?

The idea behind the expression generic application is that using a set of products in a given configuration is an application of those products. If we don't specify exactly which particular products are being used, but keep to generic notions, then we have a generic application.

In the railway systems described in the introduction there will be some kind of system that controls the individual pieces of track-side equipment that an interlocking computer governs. Such a system will typically contain a microprocessor and some kind of interface to the track-side equipment, depending on the kinds of object to be controlled.

Even without knowing which kind of microprocessor the system will be using, it will still be possible to say something about the safety properties of such a system. This will of course involve breaking the system down into smaller parts ("products") and showing that the planned configuration of products will interact in the intended way. Since you haven't (yet) identified exactly which products you will be using, you will end up identifying requirements that the products must fulfil if the application is to be safe. Given that the products have those necessary properties, you can then go on to show that the planned configuration will indeed interact in the intended way.

Here too you will have to (and be able to) say something about the quality and safety assurance activities you performed in order to guarantee that the generic application will have the desired quality and safety properties.

In real life, some of the products to be used in a generic application will be known in advance, whilst others will still be open for discussion. In our example, the track-side equipment to be controlled is probably already there, so that part of the generic application will be known in advance. On the other hand, the microprocessor to be used may still be undecided, and this will probably influence the choice of interfacing equipment. So we'll end up with a generic application that contains specific products (the track-side equipment) and "generic products" (the unspecified microprocessor etc.).

For the specifically identified products (track-side equipment in our example) there should be generic safety cases, so the safety related application conditions that those generic safety cases identify can be explicitly fulfilled in the generic application. For the "generic products" you will end up with requirements to the generic application. These will contribute to the generic application's safety related application conditions.

So here too we have enough to produce a full-scale safety case for our generic application. If we then include this "generic track-side equipment controller" in a rail traffic control system, that rail traffic control system will also be a generic application, even if everything else is explicitly identified. For the safety case of the track-side equipment, this has absolutely no effect. The fact that that generic application is being used as a "product" in a more complex system doesn't mean we have to rename the safety case!

Nevertheless, somewhere along the line we will want to build a real system, so we will have to take decisions and specifically identify which products we're actually going to use. Now our generic application becomes a specific application.

6.   Specific Application Safety Case

For a specific application life should be fairly simple. You have a generic application safety case that shows that the application will be safe if you use safe products, and you have explicitly identified products with generic safety cases that show that they are suitable for safety related applications. So all you have to do is show that those products "fit" into your generic application, i.e. that the safety related application conditions for those products are fulfilled by the generic application. You must also show that the generic application is suitable for the specific use, and finally you must show that the actual job of building the real system has been done correctly, and that operating, maintaining and ultimately decommissioning it can and will be done in a safe way.

Now this involves two fairly independent safety cases. Before you actually build and operate a specific application, the standard requires you to demonstrate that what you intend to build will be safe! That is the Design Safety Case for the specific application, and is the part where you show that the generic application you will be using is suitable and that the specific products you will be using in that generic application will result in a safe specific application.

Once you've shown that the specific application will be safe, you can start actually building it! Now you must show that the safety related application conditions that the design safety case identifies really are fulfilled, that the products that you install are the ones that the design foresaw and that the installation and integration has been performed correctly. And finally, you must show that the planned modes of operation, maintenance and decommissioning will maintain the required level of safety for the rest of the system's life cycle. This is the Implementation Safety Case for your specific application.

7. Conclusion

The preceding sections describe the ideas behind the various safety case categories that are identified in the (pre-)standard prEN 50129. The explanation given in the standard is exceptionally concise, so that there is considerable confusion about which category of safety case should be used when.

From the foregoing text the answer should be clearer: a Generic Product Safety Case should be created for "simple" products that cannot be sensibly decomposed into even simpler parts, a Generic Application Safety Case should be created for (sub-)systems whose component products are not all explicitly identified, and Specific Application Safety Cases must be created for real-life systems that are actually built and used.

The standard foresees a strong coupling between the various safety case categories, so that a large amount of the work need only be done once. Generic Product Safety Cases can be reused (referenced) in any and all safety cases for applications that use those products. A generic application safety case can be reused for all actual instantiations of that application, and a specific application design safety case can be reused for all identical instantiations of a specific application.

Used correctly, the standards can result in an exceptionally effective and economical way of demonstrating the safety of a large number of railway applications.

8. References

CENELEC (1999): EN 50126:1999
Railway Applications - The specification and demonstration of Reliability, Availability, Maintainability and Safety (RAMS)

CENELEC (2000): prEN 50129:2000
Railway Applications - Safety related electronic systems for signalling

CENELEC (2001): EN 50128:2001
Railway Applications - Communications, signalling and processing systems - Software for railway control and protection systems

SAFETY ASSESSMENT

An Assessment of Software Sneak Analysis

Authors

Graham Jolliffe BSc CEng MRAeS,
QinetiQ, MoD Boscombe Down, Salisbury, Wiltshire, UK and
Nick Moffat MA (Cantab) AMIEE
QinetiQ, Malvern Technology Centre, St Andrews Road, Malvern,
Worcestershire, UK

Abstract

In 2000 QinetiQ (then DERA) Boscombe Down were faced with results of an analysis of software for a safety critical helicopter system using a technique known as Sneak. Little was known about this technique as a Software Analysis Tool. The company responsible for the analysis, Independent Design Analyses Inc. (IDA) of Houston, Texas, was reluctant to forward details of the technique due to IPR issues.

Faced with results from a tool with unknown integrity, QinetiQ Boscombe Down proposed an evaluation of Software Sneak as a means of obtaining some reassurance of the integrity of Sneak as a software analysis tool. QinetiQ (then DERA) Malvern was asked to conduct a suitable evaluation and this paper summarises the results of the evaluation of Software Sneak Analysis. The purpose of the evaluation was to determine if Software Sneak Analysis is rigorous enough to be conducted on Safety Critical Software, and, if possible determine if it could bear comparison with better-known automated techniques.

The evaluation took the form of an experiment. QinetiQ provided suitable test code, which IDA analysed using Software Sneak Analysis. The Sneak Analysis methodology is commented on elsewhere [Brennan 2001].

IDA performed Software Sneak Analysis at their offices in Houston, Texas, on a small amount of C, Ada and Assembly code witnessed by a QinetiQ analyst. The 4 test cases, comprised approximately 1400 lines of code in total, and contained various errors, some of which were previously found by analyses performed by QinetiQ, and some of which were introduced for the purposes of the evaluation.

The software analysed during this evaluation was known to contain 22 distinct errors, considered by QinetiQ to be within the scope of the analysis (they cause the software to fail to meet the specification provided). Of these 22 errors, 19 were successfully found by Sneak Analysis and 3 were missed. In addition, IDA found 2 additional errors previously unknown to QinetiQ. 4 errors claimed by IDA are not in fact errors.

To its credit, Sneak Analysis also found 11 potential errors that are dependent on information not provided to IDA.

1.   Introduction

Software Sneak Analysis is a tool-assisted manual software analysis technique performed and developed by Independent Design Analyses Inc. (IDA) of Houston, Texas. The Boeing Company originally developed sneak Circuit Analysis in the 1960's, and software capability was developed in the late 1970's. IDA was founded in 1993 by three former Boeing engineers who had used Sneak Analysis while at Boeing and have since substantially extended the technique.

The Systems Assurance Group of QinetiQ Malvern has evaluated the technique, particularly as applied to the analysis of software (Software Sneak Analysis). The results of the evaluation are presented in this paper, starting with an outline of the technique as performed by IDA.

A Systems Assurance Group analyst representing QinetiQ performed the evaluation on site at the IDA offices. IDA engineers performed Sneak analyses of example Assembly, C and Ada code provided by QinetiQ. The QinetiQ analyst observed these analyses, and gained a first-hand understanding of IDA's Software Sneak Analysis technique, with the co-operation of IDA personnel.

The paper then discusses the evaluation, which took the form of an experiment. QinetiQ provided test code, which IDA analysed using Software Sneak Analysis. The main purpose of the evaluation was to obtain quantitative evidence of the effectiveness of Software Sneak Analysis. The Sneak Analysis methodology is commented on elsewhere [1].

The paper concludes with a summary of the results of the analyses, including a record of the effort expended by IDA when performing those analyses. The analysis results are then assessed with respect to the previously known errors in the code, and the conclusions of the evaluation are then presented.

2   Sneak Analysis - How Sneak Analysis Works

Analysis Phases

There are five phases to Software Sneak Analysis consisting of Clue List Generation, Data Preparation, Network Tree and Forest Construction, Clue Application and Discrepancy Reporting. The following paragraphs describe these phases in more detail.

Clue List Generation

In the abstract, a clue is a generic description of a system feature together with possible types of problem that may be exhibited by a system having that feature (a simple example can be paraphrased as "loops – watch out for possible infinite looping"). In practice, clues are used for finding erroneous, or at least suspect, behaviours of a system. In the case of Software Sneak Analysis, clues are used for finding software and hardware/software interface problems.

At the start of the analysis the person designated as the analysis lead prepares a list of existing clues and potential new clues. This clue list will generally contain a range of clues, varying from generic software or hardware/software interface clues through to very specific clues, perhaps applicable to particular programming languages, compilers or processors. The clue list generation phase is largely automatic – an Excel database of clues is queried, thus producing a list of those clues considered by IDA to be relevant to the software being analysed. This process appeared to be straightforward for the code analysed during the evaluation.

The clue database is the core of Sneak Analysis. This database is a major repository of experience gained by Sneak Analysis engineers within IDA and, previously, while working at Boeing. The engineers continually update and extend the clue database as their experience grows. The clue database is proprietary to IDA and confidential, so QinetiQ was not given free access to it. It was described in general terms, and a few clues were exhibited.

Data Preparation

In this analysis phase, the software components and documentation are reviewed in relation to the scope of the analysis, to determine the completeness and applicability of this data. Any missing data is requested. An inventory is made of all software components and documentation provided for the analysis; this forms the Indexed Data List (IDL). This is given to a customer prior to the analysis itself to ensure that the proper version of the data is being analysed.

Network Tree and Forest Construction

The next phase of Software Sneak Analysis is the construction of network trees and forests, which represent the original source code in a topological (diagrammatic) form. Network trees contain all the semantic information present in the original code. Their topological form makes them convenient for subsequent phases of the analysis, largely because they show the software control flow clearly. In addition, they form an electronic 'whiteboard' available for subsequent analysis phases.

Network trees and forests may be annotated during their construction, and also during clue application. Annotation is a computer aided manual process based on a documented set of procedures. The analyst responsible for annotating a particular tree or forest will study it in conjunction with any relevant documentation (either provided by the customer or obtained independently) in order to understand its intended and actual behaviour. The network tree or forest is annotated with this information and/or questions pertinent to its behaviour that the analyst feels should be answered during the analysis; these are also recorded in the analyst's Anomaly Reports / Issues List to ensure they are eventually addressed by an analyst. An automated cross-reference tool is used to generate cross-references between the network trees, variables and labels.

Clue Application

Clues from the clue list (originally from the clue database) are applied to the network trees and forests: for each clue, the analyst judges whether the system or software features referred to by the clue are present. Whenever such a feature is found in the code (in the case of Software Sneak Analysis), the possible problems cited in the clue are considered and the analyst will either convince himself that these problems do not arise or further investigate those that cannot be dismissed. The clue application phase also includes examination of the items uncovered by the automated cross-reference tool.

Discrepancy Reporting

The outputs of Sneak Analysis are reported as Sneak Condition Reports (SCRs), Design Concern Reports (DCRs), Document Discrepancy Reports (DDRs) and unresolved issues.

An SCR describes an error found by the analysis together with the circumstances in which it will occur. These errors cause failures of the system to perform as required by the specification – they are software errors in the case of Software Sneak Analysis. IDA refers to these errors as 'sneaks', and the circumstances in which they occur as 'sneak conditions'. (Hence the name "Sneak Condition Report".)

DCRs describe 'design concerns': apparent problems with the (software) design, given the intended use of the software.

DDRs describe apparent inconsistencies between documents that accompany the code.

Possible problems are reported at intervals during Sneak Analysis, categorised as above, or perhaps just flagged as unresolved issues (suspect or of concern). This is intended to give the customer early visibility of possible problems. It also provides an opportunity for feedback from the customer, which may help with the remainder of the analysis. All such problems still considered genuine on completion of the analysis are documented in the Sneak Analysis final report (up to that point they remain provisional).

Recommended Fixes

In addition to the various problems found by Sneak Analysis, the final report suggests accompanying remedies. These remedies might be general or very specific, often constituting an alternative section of code that, it is claimed, avoids the identified problem. These recommendations are offered in an attempt to provide the customer a sneak-free solution to the identified problem.

Sneak Analysis - Quality Assurance

Quality assurance activities are performed throughout Sneak Analysis and appear to be thorough.

Typically, the analysis lead is responsible for co-ordinating QA activities throughout an analysis. Pre-defined procedures exist for analysts to check code in and out of a centrally controlled electronic repository. Control is maintained via restriction of electronic file access permissions. Validation activities, including peer review, are also mandated.

Access

At the outset of the evaluation, an IDA analyst presented an overview of Sneak Analysis to the Malvern analyst. This presentation focused on Software Sneak Analysis. In addition, the Malvern analyst discussed aspects of Sneak Analysis with IDA analysts throughout the evaluation, often interrupting their analyses to enquire about their activities.

However, the Malvern analyst was not allowed free access to the clue database. It was described in general terms, and a few example clues were shown and described.

It had originally been planned that IDA would reveal the particular clues that enabled Sneak Analysis to find the issues raised. It was later agreed that IDA would provide this information only for a few selected issues chosen by QinetiQ; documentation of the large number of issues and reportable items would have taken up too much analysis time. However, ultimately it was decided that this information would not be required at all, since there was no reason to doubt IDA's claim that all issues raised were obtained by clue application.

3 Evaluation

The four test cases (comprising approximately 1400 lines of code in total) were analysed by IDA at their offices in Houston, with the Malvern analyst present. The analysis lasted 2.5 weeks, involving up to four IDA analysts simultaneously.

The Malvern analyst was present during the Sneak analyses in order to (a) answer questions about the test code (acting as a customer for the IDA analysts), and (b) observe the progress of the Sneak Analysis and ask questions, in order to gain a good understanding of Software Sneak Analysis.

At the start of the evaluation, IDA were not told how many errors were known to exist in the code, nor even that all code contained errors. However, it was clear to IDA that there were likely to be errors in some of the code – otherwise it would have been inappropriate code for the evaluation. In fact, all test cases contained at least one known error.

Source Code

Test Case 1: Assembler

A small amount of Motorola 68020 assembly code (together with co-processor instructions) was chosen for the first test case. It consists of 27 declarations/instructions. This code is a small fragment derived from code that currently runs on a real-time system, and into which QinetiQ introduced an error for the purposes of this evaluation. Although small, this test case employs quite a subtle coding trick.

Test Case 2: C

Test Case 2 consists of 330 non-blank, non-comment lines of C. It can exhibit a range of run-time errors, including division by zero and infinite looping. In fact, it is derived from demonstration code for the automatic exception analysis tool, Merle, being developed within the Systems Assurance Group of QinetiQ Malvern.

Test Case 3: C

The C code chosen for the third test case is derived from demonstration code distributed with the PolySpace Verifier, which is an exception analysis tool developed by PolySpace Technologies. The original demonstration code is copyright PolySpace Technologies. The test case derived from that code consists of 480 non-blank, non-comment lines.

Test Case 4: Ada

The Ada code chosen for the evaluation forms test case 4. It consists of 545 non-blank, non-comment lines of code. This code was manually edited from software generated automatically by Simulink[1] from a Simulink control law diagram. The manual editing improved the code structure with the intention of preserving its semantics, but also deliberately introduced errors for the purposes of this evaluation.

Summary of Code

In total, there were approximately 1400 non-blank, non-comment lines of code (850 lines, excluding C header files and Ada package specifications). The code and in particular the errors was chosen to be representative both in terms of functionality and the types of error that could typically be expected to occur.

Specification of Required Behaviour

Test Case 1: Assembly

QinetiQ defined the required behaviour of the assembly code to be the functional behaviour claimed in the comment at the top of the assembly source file, together with the requirement that run-time exceptions should not occur. The assembly source code description states that the code implements straight-line interpolation.

---

[1] Simulink is a component of Matlab, a tool marketed by MathWorks Inc.

Test Cases 2 and 3: C

The sole requirement defined by QinetiQ for the C code was that it should not exhibit run-time exceptions.

Test Case 4: Ada

A Simulink diagram was provided. The requirement defined by QinetiQ was that the code should correctly implement the control law specified by the Simulink diagram, and no run-time errors should occur.

Authorised Assumptions for Sneak Analysis

For all test cases, the IDA analyst was free to ask the QinetiQ Malvern analyst of this paper for information that a true customer could credibly supply. The Malvern analyst fielded these questions with the assistance of colleagues at QinetiQ Malvern. Each such assumption made by the IDA analysts was recorded in an Open Reports / Issues List. Additionally, the analyst for test case 4 was asked to assume that large portions of the code (corresponding to duplicated components of the Simulink control law) were identical, up to necessary name changes for variables, constants and procedures. This assumption had been verified by the Malvern analyst, in order to allow IDA to concentrate on analysing only one instance of this duplicated code.

Analysis Outputs

The usual reports from each analysis were provided as listed below. It was mutually agreed that formal reports and the final report were not required for the evaluation.

i.   Analysis schedule (a record of progression through the analysis phases)

ii.   Indexed Document List (IDL)

iii.   Program Operations Diagram (a POD shows the relationship of the network trees or forests to each other)

iv.   Open Reports / Issues List

In addition to the outputs normally provided to each customer as standard, additional internal documentation was provided to assist the evaluation. This documentation was produced as part of IDA's analysis process and was used to track progress as well as quality control. The additional documentation for each analysis included:

v.   Software listing

vi.   Network trees (topological representation of code) with cross references table (the cross references track declaration, update and use of all variables)

vii.   Annotations for a particular network tree (NT)

viii.   Network tree completion log (tabular record of NT completion)

ix.   Network tree schedule (number of NTs completed plotted against time)

x.   Clue application completion log (tabular record of NT completion)

xi.   Clue application schedule (number of clues applied plotted against time)

4 Results of the Evaluation

Open Report / Issues Lists

The Sneak Analysis results ("issues") reported in the Open Report / Issues Lists were reproduced in full in the format of result tables. Those result tables also show QinetiQ's judgement of each issue with respect to the correctness of IDA's reasoning and the issue category. Although, the result tables are not provided in this paper, the judgements are explained below, together with summary tables showing the distribution of issues according to correctness of reasoning and issue category.

Correctness of Reasoning

IDA's reasoning was found to be completely correct, sometimes partially correct, and sometimes wrong. In two cases, the reasoning is not clear enough to judge. Also, for some issues (e.g., assumptions) judgement of correctness is not appropriate.

Table 1 shows the correctness of all the issues reported by Sneak Analysis. These issues are produced as a result of applying the clue lists and so, contain real errors and non-errors (these are further categorised below). Inevitably, the number of issues will outnumber the actual number of errors present. IDA then applies their reasoning to determine if the issues need to be addressed. Duplicated issues are only counted once. Separate issues reported together count separately and issues that were judged as inappropriate are listed as Not Applicable.

	Test Case 1	Test Case 2	Test Case 3	Test Case 4	Total
Correct	7	16	23	28	74
Partially Correct	3	1	3	2	9
Incorrect	0	1	9	1	11
Unclear	2	0	0	0	2
Not Applicable	3	0	0	2	5
Total	15	18	35	33	101

Table 1: Correctness of Sneak Analysis Results

Issue Categories

In order to assess the effectiveness of the analysis it is necessary to categorise the errors reported by Sneak analysis. The first two issue categories are:

- **Not errors.**
This category represents those issues that are not errors. All notes, comments, assumptions, and advisories are counted here, but also design concerns that are not errors. For example, this includes unreachable code, portability and maintenance issues.

- **Potential errors.**

This category includes those issues that IDA claimed could potentially make the software unfit for purpose, depending on out of scope information separate from the code and software specification provided. (IDA typically classifies such issue reports as DCRs.)

The remaining categories refer to errors that are within the scope of the evaluation (they do not depend on out of scope information):

- **Falsely claimed errors.**

These are the in scope errors wrongly claimed by IDA. (Contrary to IDA's claims, they do not cause the software to fail to meet its specification.)

- **Known errors.**

These are the in scope errors found by Sneak Analysis that were previously known by QinetiQ. Most had been found by analyses performed using one of QinetiQ's analysis tools (Merle) or a competing tool (the PolySpace Verifier), but some had been deliberately inserted into the code for the purposes of this evaluation. The number of known errors prior to analysis was 22. However, 2 of the errors in Test Case 1 are duplicates and have not been counted twice. Consequently, the total number of different errors known to QinetiQ is 20.

- **New errors.**

These are in scope errors found by Sneak Analysis that QinetiQ were not previously aware of.

Table 2 shows the distribution of issues among the issue categories and also the number of in scope errors missed by each Sneak Analysis. The difference in totals is due to the missed errors.

		Test Case 1	Test Case 2	Test Case 3	Test Case 4	Total
Not errors		13	10	22	22	67
Potential errors		0	5	3	3	11
In Scope Errors	**Falsely claimed**	0	1	3	0	4
	Known	2	2	7	6	17
	New	0	0	0	2	2
	Missed	0	0	2	1	3
Total		15	18	37	34	104

Table 2: Categorised Sneak Analysis Results

Assessment of Analysis Results

Correctness

We focus first on correctness: are the issues claimed by Sneak Analysis genuine, and are the supporting arguments correct?

First, consider the proportion of claimed errors that are genuine. Table 2 shows that 19 (Known and New) of the 23 (Known, New and Falsely Claimed) in scope errors claimed by IDA are indeed genuine.

Next, consider the correctness of the reasoning used by IDA in support of the issues reported by Sneak Analysis. Table 1 shows that, of the 96 arguments supporting issues raised by Sneak Analysis, 74 are correct, 9 are partially correct, 11 are incorrect, and 2 are unclear.

Coverage

The most critical statistics, of course, relate to coverage and determining how effective is Sneak Analysis at finding genuine errors.

Table 2 shows that Sneak Analysis found 19 of the 22 genuine in scope errors. By this simple statistic, Sneak Analysis does seem effective at finding most serious problems. The 3 missed errors demonstrate, however, that it is not an infallible technique.

A claimed strength of Sneak Analysis is its ability to discover a wide range of error types. The fairly large number of potential errors that were identified as shown in table 2 provides evidence for this. In addition, several maintenance and portability issues were found by Sneak Analysis – these are included (together with other issues) in the "Not errors" category of table 2.

Record of Analysis Time

Resources Used

This section presents IDA's time spent analysing the test cases. The Malvern analyst concurs with these measurements.

The activities of the analysis lead included: Initial Customer Review (6 hours), Progress Reviews and Sneak Analysis Discussions (16 hours), Chart Generation (8 hours) and Quality Control (50 hours). In addition 15 hours were spent assisting the analyst with Test Case 1 – this time is included in the 67 hours of Clue Application shown in Table 3.

The time required for clue list generation was negligible. This may not generally be the case for new or novel languages that IDA have never analysed – effort would be required to generate language-specific clues and add them to the clue list if the analysis is to find language-specific issues. The significance of this depends on the likelihood of language-specific issues; there might be none.

Table 3 also shows the time declared by IDA for Progress Reviews and Sneak Analysis Discussions and for the analysis phases Data Preparation, Network Tree

Construction and Clue Application (the formal Discrepancy Reporting phase did not occur). As well as IDA's declared times, the table also shows the number of lines of code for each test case and the average number of lines of code analysed per day. Note that these figures do not take account of the overhead activities of the lead analyst. This would require an estimate of how that overhead was split among the four test cases, not attempted here.

Test Case 1 contained only one error, yet was challenging because of the subtle nature of the code. This had a clear effect on the lines of code per day analysed by Sneak Analysis: 2.4 is a very low figure. Sneak analysis of Test Case 1 would certainly have been faster if the input format had been revealed to the IDA analyst. The customer at start of an analysis or upon request generally provides input format. The majority of IDA's time analysing this test case appeared to be spent trying to deduce the input format. IDA's quoted historical average rate is 40 lines of assembly code per person per day, which includes all analysis phases.

	Test Case 1	Test Case 2	Test Case 3	Test Case 4
Progress Reviews and Sneak Analysis Discussions	4 hours	5 hours	5 hours	11 hours
Data Preparation	8 hours	2 hours	3 hours	12 hours
Network Tree Construction	8 hours	12 hours	25 hours	31 hours
Clue Application	67 hours	23 hours	31 hours	48 hours
Total (excluding progress reviews and discussions)	83 hours	37 hours	59 hours	91 hours
Total Lines of Code	27	330	480	545
Lines of Code per Day	2.4	67	61	45

Table 3: Time Spent Performing Sneak Analysis

The remaining test cases were analysed at a rate between 45 and 67 lines of code per day. It should be remembered that these figures do not include effort that would normally be spent during the Discrepancy Reporting phase or the (less significant) overheads of quality assurance activities, progress reviews and chart generation. According to IDA's historical records, these items generally take a total of between 10% and 20% of the analysis resources.

Mitigating Factors

IDA requested that mitigating factors be considered in this evaluation paper. In particular, the claimed mitigating factors relate to analysis overhead, concurrency of analysis and artificiality of the code. The following paragraphs alternately give a paraphrased summary of the mitigation claimed by IDA (italic font), and an assessment of the validity of these claims (plain font).

IDA: The resource figures shown include a disproportionately large overhead, due to the evaluation format.

Response: The Initial Customer Review is a clear overhead which is required at most once, whatever the size of the analysis task. Progress reviews and Sneak Analysis discussions are also overheads. Progress reviews during the evaluation were very frequent, averaging one every two days of analysis; IDA have stated that this is unusually often, since reviews would normally occur monthly or less frequently. Sneak Analysis discussions were also unusually time-consuming and can be disregarded by this evaluation – analysis would not normally be performed as part of an evaluation. So, the overhead was indeed disproportionately large for progress reviews and Sneak Analysis discussions. Even so these overheads are not included in Table 3.

IDA: Larger analysis tasks would involve more concurrent activity. This would enable more efficient use of the available man-hours and provide greater opportunities for discussion, which would in turn enable problems to be resolved more efficiently. The use of network trees allows larger analyses to be handled as easily as smaller analyses because network trees partition the system into the same manageable sizes in either case.

Response: It is well known that performing tasks in parallel can introduce prohibitive communication and co-ordination overheads [Brookes 1995]. The sneak analyses involved almost no concurrency of the type described, so more concurrent activity might indeed have helped these analyses. However, larger analysis tasks tend to be more complex, and tools tend to have a greater capacity for coping with complexity than humans have. IDA's claim that network-trees partition large systems into manageable sizes is fair, but this can only mitigate against some of their complexity.

IDA: The code used for this evaluation is artificial, containing disproportionately many issues and so increasing analysis time. Software Sneak Analysis historical records (based on millions of lines of code over 20 years) show an average of 5 reportable items per 1000 lines of code. This evaluation shows a rate of 70 reportable items per 1000 lines of code (based on the 22 errors discussed above the rate is approximately 17 reportable items per 1000 lines of code).

Response: This really only applies to Test Case 3 which, it is conceded, contains more issues than would be expected on average. It is a fair argument for perhaps considering analysis time per issue, or per error, rather than analysis time alone.

5   Discussion of Results

Sneak Analysis found a broad range of errors, including many considered out of scope of the evaluation (they depend on information that was not available to IDA). This is certainly a strength of Sneak Analysis.

The Sneak Analysis of Test Case 4 found two errors previously unknown to QinetiQ. It is certainly to the credit of Sneak Analysis that these two errors were found during this evaluation. Sneak Analysis has highlighted a known shortcoming of QinetiQ's original analysis technique, which has since been properly addressed by a more sophisticated QinetiQ analysis technique. (The original technique's shortcoming was known and documented at the time of the original analysis.)

The evidence available – general discussions with IDA analysts and the example clues seen – suggests that the clue database for Software Sneak Analysis is essentially a code walkthrough checklist, structured to enable convenient generation of clues applicable to particular types of system.

Without free access to the clue database, it is not possible to directly assess the level of expertise required for clue application, or the completeness of the clue database. This paper can only judge those factors based on the Sneak Analysis results obtained for the evaluation software and on the Malvern analyst's observations during the analyses. On that basis, the level of expertise required for Sneak Analysis is substantial. IDA's clue database is a significant commercial and technical asset.

Sneak Analysis is a good manual technique, perhaps exceeding other manual static techniques such as Fagan Inspection. It is systematic and appears to be well controlled. Its quality assurance activities appear to be rigorous and appropriate. Issue reports are given to the customer at regular intervals in advance of the final report – a useful feature.

Sneak Analysis can reasonably be expected to find a large proportion of genuine errors. However, it should not be relied on to find all possible errors. Experience with manual techniques has shown that they are in general prone to human error, and Sneak Analysis is no exception (evidenced by the three missed errors).

From QinetiQ Boscombe Down's perspective, this result only partially satisfied the question of integrity for this technique. For the assessment of the safety critical system in question QinetiQ Boscombe Down were also interested in the range of errors that Sneak Analysis can find. To this end, separately from the evaluation report, the Malvern analyst responsible for the assessment identified which tools and techniques could have been used to find errors of the type identified by Software Sneak Analysis. Whilst there were some simple errors that could be detected by a compiler, many others would have necessitated the use of more powerful analysis tools such as MALPAS semantic analysis. In some cases compliance analysis would be necessary. So whilst Sneak is manual process and therefore prone to error, it does bear comparison with more powerful automated techniques. IDA also found a possible error that would otherwise have required manual code reading and knowledge of the hardware. By being able to provide analysis of the software/hardware interface, Software Sneak Analysis might be unique.

This paper has not addressed the question of value for money of Sneak Analysis; that can be judged by reference to the resources used and the results obtained.

An important consideration is the scalability of Sneak Analysis. The dependence on skilled analysts imposes a limit on the practical scalability of Sneak Analysis, though this is hard to estimate. More fundamentally, the most skilled and labour intensive manual activity of Sneak Analysis (Clue Application) appears to scale linearly with the amount of code. This is consistent with IDA's time and cost quotes [Malone 2001], but contrasts with automatic techniques for which, in QinetiQ's experience, associated manual activities can scale better than linearly.

6 Conclusion

Having performed this evaluation, the QinetiQ Malvern System Assurance Group's view is that Software Sneak Analysis is a worthwhile technique. However, in common with other manual techniques, it does not provide the same level of assurance as formal techniques, due to human error and the reliance on human experience.

As long as it remains manually intensive, some doubts over the use of Sneak in a software context will inevitably remain. However, the result of this evaluation has been a key building block in the development a safety argument of the safety critical system in question. That in turn will enable QinetiQ Boscombe Down to provide positive recommendations for this system's use in service after many years of doubt over the system's integrity.

References

[Brennan 2001] "An Assessment of Sneak Circuit Analysis and its potential for the provision of evidence in support of software integrity", Draft A, 17th February 2001.

[Malone 2001] Letter with subject "Sneak Analysis Evaluation" from Mr Mitch Malone, Vice President of IDA, Inc. to Mr Graham Jolliffe, Technical Leader for Software Assessment at QinetiQ, Boscombe Down, dated 12th January 2001.

[Brookes 1995] "The Mythical Man Month", 2nd Edition.

Processes for Successful Safety Management in Acquisition

RF Howlett; Land Systems Safety Office; United Kingdom Ministry of Defence

Abstract

Over the last few years UK MOD has adopted a risk based Safety Case approach, as defined in UK Defence Standard 00-56 reference 1, as the means for demonstrating equipment is safe for its purpose and as the basis for establishing the through-life Safety Management Systems for the equipment. Within the author's area of work, Land Systems, the Safety Case is developed in three stages. These stages (setting the requirements, validating the design, and qualifying equipment for service) combine the assessment, risk reduction and through-life safety management activities for a project. The process developed is flexible enough to enable it to be tailored to accommodate a wide range of diverse equipment, and procurement strategies.

During the last year, to assist Project Teams in the management of safety and improve equipment safety assurance within the UK MOD work has been undertaken to develop a series of process maps. These chart the safety management and assessment requirements across the SMART Acquisition philosophy followed by the UK MOD. The work has mapped out the steps to be taken and deliverables to be provided at each stage of the acquisition cycle. This paper presents the maps developed and describes the results of the work in generic terms.

In the future it is intended that the maps will be used to develop metrics to measure the performance, refine the safety management process, assist in training and developing competencies, therefore providing continued assurance as to the overall quality of work being undertaken.

1 Introduction

Over many years the UK MOD has progressively adopted the Safety Case regime to demonstrate that equipment is of safe design and is suitable for its service role. The Safety Case structure followed is used to set requirements, feeding those in to the requirements process, demonstrate the design has achieved those requirements and reduced risk to a tolerable level. It is also used to develop and implement a management system that ensures effective management and maintenance of safety performance through life.

Successful Project Management requires not only the effective management of a number of technical areas, which includes safety, but also the control of programme risk. The approach adopted for Land Systems uses a defined three stage approach to developing and delivering safety and matching this to the acquisition process, minimises project risk.

The effective management of any situation requires understanding of the issues and ensuring that the resources are available to deal with them. To this end it is vital that an organisation has procedures, competent people and mechanisms in place to enable it to respond to its responsibilities and achieve its objectives. This is particularly important in the area of safety management as any organisation has legal obligations as well as business objectives to meet. Within defence a balance of safety risk and operational performance is also needed to ensure defence capability is maintained.

2   Safety Management Arrangements within Land Systems Project Teams

The requirement for the management of safety for Land Systems equipment is set out in internal UK MOD procedures. These procedures require all Project Teams managing Land Systems equipment to establish a through-life Safety Management System (SMS) for the projects they manage. The interaction of the key elements of the SMS is shown in the Figure 1. These elements of the SMS are listed below:

- **Safety Panel** operates the Safety Management System, setting requirements, manages the Safety Plan, carries out reviews and audits of the SMS. Membership comprises representation from all authorities that exercise responsibility for the project throughout its life (e.g. Project Manager, Designer, Operator, Maintainer, and others). They set the Safety Requirements, accept the Safety Case, and manage the Safety Case in service. In some instances the Safety Case is reviewed by an body independent of the project management and Safety Panel.

- **Safety Plan** documents the Safety Management arrangements for the project, includes system boundaries, scope of the SMS, objectives, responsibilities, requirements and the safety programme for the project. The Safety Plan forms part of the project's Through Life Management Plan.

- **Safety Case** is based on an assessment of the equipment/system being developed. It provides a justification, through evidence and analysis that the equipment is safe for its given purpose. It also sets out the means and requirements for maintaining and operating the equipment safely throughout its life, and demonstrates that it complies with all relevant legislation.

- **Hazard Log** is the core of the Safety Case. It is used to record the assessment results, decisions and safety related information, such as defect and accident reports. For Land Systems projects a database application 'Cassandra', see reference 2, has been developed as a standard tool, which is also being adopted by other areas in the UK MOD and industry.

Figure 1 - Land Systems Safety Management Model

3 Elements of the Safety Case

Within Land Systems the Safety Case is divided into three parts:

- The first part is used to generate the criteria against which the performance and acceptance of the project will be measured and it sets the standard that the designer is required to achieve. It is initiated, as part of the development process for the specification, during the Concept phase of a project. As the design matures, this part of the Safety Case is updated and used to review the overall status of the equipment. The objectives and requirements are derived from a safety risk assessment of the capability or concept being developed.

- The second part is generated progressively as the equipment/system is being designed and developed, through the assessment and demonstration stages of the project. Its purpose is to demonstrate progressively that the project is meeting the requirements set out in the first part of the Safety Case. This part is normally generated and produced by the designer, who justifies that the equipment is safe and suitable for service through the presentation of appropriate evidence.

- The Project Team prior to acceptance of the equipment produces the final part of the Safety Case, the Operational Safety Statement. It is used to demonstrate that the arrangements to manage and control residual safety risks exist and are adequate, or are being put in place. This part should demonstrate that all the safety management arrangements enable the operators and support authorities to control residual risks, and maintain the required levels of safety through life.

In effect the Safety Case comprises of a series of reports delivered as the equipment project builds through the Acquisition Cycle described below. This approach ensures progressive achievement of safe design that meets the requirements and hence projects goals.

4   The Acquisition Cycle

The Acquisition Cycle, requires delivery of a number of key elements through the various stages on a projects life cycle. These key project elements are described below for each of the phases of the Acquisition Cycle:

- **Concept Phase** – During this phase, based on an operational capability derived by the armed forces a series of requirements are defined and are used to populate a User Requirements Document (URD). Various technical solutions and procurement strategies are explored with industry, boundaries are set for the project and a business case made. When potential solutions are identified a business case is made to support the continuation of the project. If the business case is successful Initial Gate approval is given and further funding to develop the project into the Assessment Phase is released.

- **Assessment Phase** – Building on the work undertaken in the Concept Phase a System Requirements Document (SRD) is developed from the

URD and the investigation and assessment of various technical and procurement solutions. When cost-effective solutions are identified which meet the user requirement within the specified project boundaries a Main Gate submission is made for approval and main funding for the project.

- **Demonstration Phase** – During this phase the technical solutions are evaluated and supplier identified for the production of the agreed solution.

- **Manufacture Phase** – Undertake production of the agreed solution, confirming that the delivered solution still meets the requirement and conduct System Acceptance and field the equipment. Where appropriate, transfer management of the project to the in-service support authority.

- **In-Service Phase** – Maintain the system for operational use, revising and developing the requirement through modifications and improvements, to ensure cost-effective ownership and continuing operational effectiveness.

- **Disposal Phase** – Ensure the efficient, effective and safe disposal of equipment. Disposal needs to be managed through out the life of a project, from the disposal of research and trials equipment, through damaged and defective equipment through to end of life disposal issues.

The relationship between the Acquisition Cycle and the elements of Safety Case, described in 3 above, and key project milestones are shown in Figure 2 below:

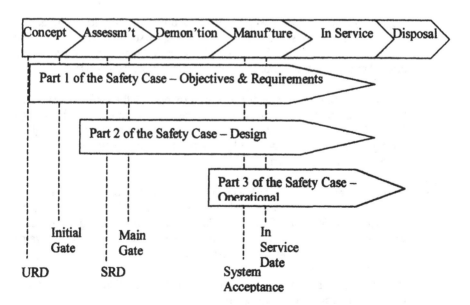

Figure 2 – Relationship between Acquisition Cycle and the
Safety Case

5 The Process Model – Initial Work

Initial work started on the production of a process model with the development of a top-level model that mapped the primary elements. The top-level diagram, see Figure 3, shows activities divided into five primary elements:

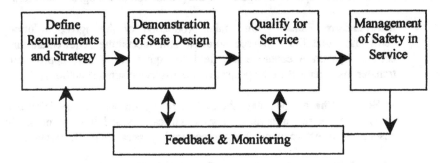

Figure 3 - Top Level Safety Process Model

- **Defining Safety Requirements and Strategy** – equates to the first part of the Safety Case. The Safety Panel would be formed at this time to carry out this work and publish the Safety Plan and commission work to develop the requirements part of the Safety Case.

- **Demonstration of Safe Design** – equates to the second part of the Safety Case. This demonstrates that the designer/supplier has met the requirements, also identifies instructions and/or management provisions needed to support the equipment with any limitations on operation.

- **Qualify for Service** – equates to the third part of the Safety Case. This brings together any current safety management regimes and the design Safety Case to demonstrate that all the required management and logistic arrangements are in place and working to enable safe and sustained operation of the equipment.

- **Management of Safety in Service** – through the Safety Management System, the Safety Panel maintain the Safety Case. This is undertaken by monitoring of feedback from the field and changes (role, legislation etc.) that affect the system. Appropriate modification may need to be instigated either to the design or operating procedures.

- **Feedback & Monitoring** - The top-level model identified several feedback routes. They relate to feedback into the design from incidents in the field and changes to role etc. and the capture of experience from the equipment to inform decisions when planning future or similar systems.

6 The Process Model – Maps

The following set of Figures show the key elements of the safety management and assessment process for each phase of the Acquisition Cycle. The majority of the activities shown also have with them 'child processes', which are not shown or presented in this paper. Figure 4 shows the symbols that have been used in the maps:

Figure 4 – Process Map Symbols

Concept Phase - Work on the Safety Case is initiated during the Concept Phase, see Figure 5, with the generation of Part 1 of the Safety Case. This is used to determine the safety objectives and requirements for the project. This is done through a high level assessment of the capability requirement, which determines potential hazards and is used to identify the regulatory regime/s with which the project may need to comply. The output from the first part of the Safety Case is then used to populate the outline specification with safety requirements, examine various options and support the initial business case for project approval.

During this important phase work is initiated with the formation of the Safety Panel, comprising all the stakeholders that will have responsibility for the project through life. As the eventual authority for the safety of the equipment being procured the Safety Panel monitors, scrutinises the work, and establishes the Safety Management System for the project.

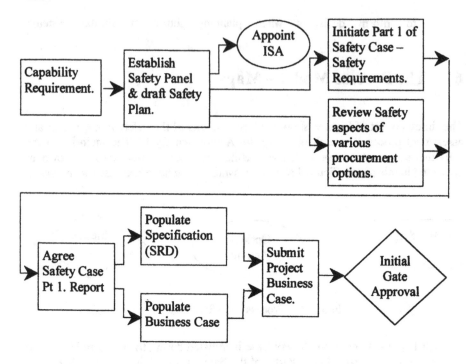

Figure 5 - Concept Phase

Assessment & Development Phases - During the Assessment and Development phases, see Figures 6 and 7, the first part of the Safety Case is refined in line with the options being explored with its output being used to support procurement and business strategies and to refine the SRD. In parallel the second part of the Safety Case is developed where there may be several separate Safety Cases being developed depending on the number of solutions being explored. This second part of the Safety Case is used to inform and refine the design and demonstrate how the requirements, set out in the first part of the Safety Case, are being met.

Throughout these phases the Safety Panel will continue to monitor and scrutinise the work, endorsing the Safety Programme and Case. The safety work will be used to and support the project's Business Case at Main Gate and the evolving acquisition strategy during the Assessment phase, and the decision to start production at the end of the Demonstration phase.

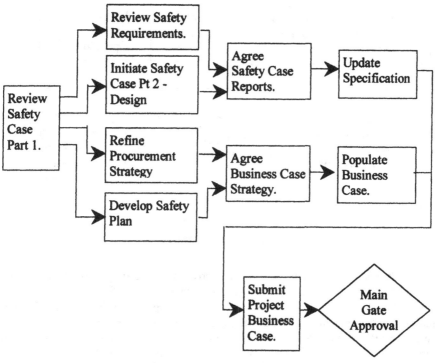

Figure 6 - Assessment Phase

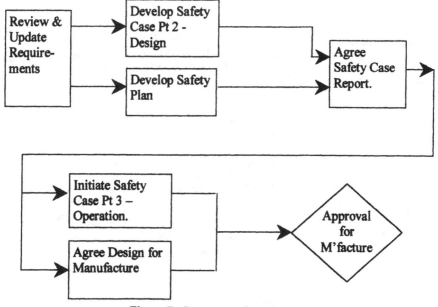

Figure 7 - Demonstration Phase

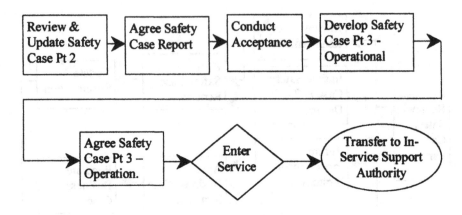

Figure 8 - Manufacture Phase

Manufacturing Phase - The primary objective of the Safety Case work in the Manufacturing phase is to support System Acceptance, see Figure 8. The second part of the Case is used by the Design Authority/Supplier to demonstrate that the final design and any changes made during production still meet the requirements set out in the SRD and the first part of the Safety Case. The Project Team initiates the third and final part of the Safety Case, the Operational Safety Statement, during this phase. This final part of the Safety Case justifies that arrangements exist, or are being put in place, to ensure that the measures for controlling risks are adequate and will sustain safe operation in service.

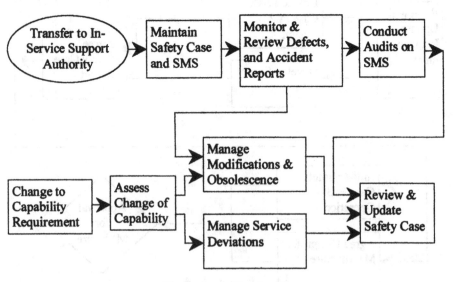

Figure 9 – In-Service Phase

In-Service Phase – During the In-Service phase, see Figure 9, once equipment has entered service the primary task of the Safety Panel and the SMS are to maintain the Safety Case. These ensure that all control measures and arrangements identified and claimed by the Safety Case are maintained for the continued safe operation of the equipment. This is done by monitoring safety performance, auditing safety arrangements and conducting reviews and updates of the Safety Case as conditions change through life. The baseline set by the requirements in the first part of the Case, coupled with the design part of the Case are used to evaluate changes and modify, if necessary, the arrangements set out in the operational part of the Case, and to provide feedback to future projects.

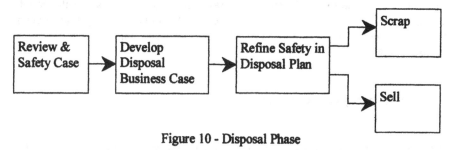

Figure 10 - Disposal Phase

Disposal Phase – The requirements for disposal of the equipment should have been addressed throughout the generation and development of the Safety Case. It is still necessary, however, during the Disposal phase to review the Case when planning disposal to ensure that all the appropriate conditions are being met, see Figure 10. In the case of an installation or facility it may be necessary to generate a specific Safety Case for disposal or demolition.

7 Discussion

Mapping the processes was undertaken in conjunction with several Project Teams, this enabled their experiences and needs to be addressed alongside those of the policy office that produce the instructions and procedures. These processes therefore combine the practical view and needs of the Project Teams with the need to provide safety assurance at a corporate level.

The mapping process has also clarified the interaction between the safety assurance process and the Acquisition Cycle. It is drawing Project Teams' attention to the need to concentrate effort in defining the requirements, particularly the impact that these might have later in the programme if they are not fully developed. The Project Teams found it particularly useful to put a timetable to the current published procedures because this would improve early planning. Through the use of process maps, teams will be more readily able to identify deliverables, reports and when they are required, with the aim of reducing programme risk.

Capturing the 'Land Systems' process has already identified the differences that exist between this and other sectors within the UK MOD. As other sectors map their processes, by comparison of the outcomes should ensure consistency in the approach to overall policy for system safety. It should then be possible to build on strengths in common areas.

There is a danger associated with the publication of the maps. Project Teams could see them as a 'tick in the box' approach to safety management, or as a substitute to training staff. Effort will be needed to ensure they complement safety management and the assurance process, and that Project Teams take due consideration of what is required in their particular situation.

8   Future Development

Having established an agreed model, the work it will be used in the future as the basis for the development of procedures, in addition to use by Project Teams as a vehicle of best practice. The model could also be used for developing and delivering training, as well as tools and standards for safety management.

9   <u>Conclusion</u>

The work, to date, provides a model that will benefit Project Teams and others involved in the management of safety for Land Systems equipment. The maps combine the corporate policy and procedures and programme experience. The model along with the documents it references clearly identifies the roles and importantly tasks to be undertaken at each phase in the Acquisition Cycle. This leads to a more focused approach that will greatly aid the management and demonstration of safety. In the longer term the work provides a strong basis from which to evolve and develop good practice, training and procedures.

References

1. UK MOD, Defence Standard 00-56, Safety Management Requirements for Defence Systems, December 1996.

2. Howlett, R.F., Improving Hazard Management Through Life, ISSC, 2001.

Assurance of safety-related applications on a COTS platform

Dr Clive H Pygott

System Assurance Group, QinetiQ(Malvern) UK

Abstract

Many computer-based information systems act in an advisory manner, where an obvious failure can be tolerated, but a 'plausible but wrong' output is hazardous. It is also often necessary to support such safety-related applications on a COTS platform, as to the user it is simply another task to be performed and be used alongside other office automation tools.

This paper looks at how 'fail-safe' applications can be implemented on a COTS platform with appropriate levels of assurance. The approach is based on monitoring software running in parallel with the actual application, providing a 'sanity check' on the outputs displayed. A significant issue is how to avoid 'common cause' or latent failures, such as the operating system failing to run the monitor. The approach is illustrated by a number of case studies: a military pilot's planning aide, an 'ATC-like' display and a critical document control.

1 Introduction

1.1 Background

This paper considers those safety-related applications which need to be run on a COTS platform, because to the user it is simply another office automation tool to be used alongside word-processors, spreadsheets, etc. The use of two machines, one general purpose, the other for the safety-related applications is not acceptable in terms of cost, space, etc.

A significant factor for all the applications considered here is that they are fail-safe, in the sense that an obvious failure of the system is not a hazard, as the user has sufficient time to achieve the same objective by some other means. However, the user acting on 'plausible but wrong' output would be hazardous.

Historically there are three approaches to the justification of COTS in safety related systems (Bishop, Bloomfield & Froome 2001) (Jones et al. 2001) (Scott, Preckshot & Gallagher 1995) (McDermid 1998):

- get more information
- limiting the demands made on the system
- limiting the authority of the system

A number of the referenced reports consider the sources of additional information that may be used to justify COTS. The essential problem with this approach is that there is no way of determining how much information is 'enough'. That is, there is no agreed calculus for establishing and combining the value of different pieces of evidence.

Encapsulation, in the sense of limiting the demands made on the COTS has two main problems, firstly it is not easy to see how a firewall can be placed between an application and the operating system that supports it, and secondly the idea of restricting demands to some core and, by implication, trusted subset has no real rigorous foundation. Why should that subset be more trustworthy than the rest?

This paper, therefore, concentrates on limiting the authority of the system. The techniques described aim to ensure that all errors lead to obvious failure and so are not hazardous.

1.2 Approach

The work reported here is based on an MOD funded project, so is looking to satisfy MOD's safety requirements (*Defence Standard 00-56* 1996) (*Defence Standard 00-58* 2000), which are broadly IEC61508 compliant. The work is more fully reported in (Pygott 2002).

The approach adopted uses a model of an operating system, in terms of the services it provides. Hazop analysis is then used to identify how failures in those services could manifest. This leads to derived safety requirements for the application, and requirements for monitoring processes to identify and report errors.

It is worth noting that the Hazop is performed on a very high-level operating system model, but the justification of particular safety methods may require more detailed knowledge of the software (and possibly hardware) architecture. For example, where an application writes data to the file store, the requirement for this activity alone may identify hazards associated with failure to write or corruption of the data. However, when the protection mechanisms are considered, issues such as the location of caches become important, as checking for correct writing by reading and confirming the data just written provides no benefit if that data is taken from a cache rather than the file store.

The justification of safety related applications on COTS is illustrated by three case studies.

2 Case study 1: An Electronic Operating Data Manual

2.1 Introduction

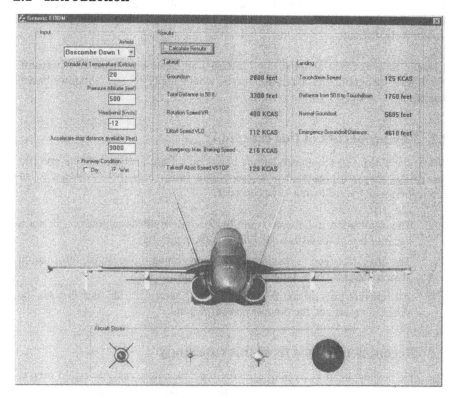

Figure 1: An illustrative EODM program

An Operating Data Manual (ODM) is an aircraft performance document that shows how performance parameters, such as required runway length, depend on factors such as meteorological conditions, drag, mass etc. Rather than use the printed ODM, pilots prefer an Electronic Operating Data Manual (EODM) as illustrated in Figure 1; a computer program that performs the ODM calculations for them (this image is only illustrative, if the text is illegible it does not affect the argument in the rest of the paper). In safety assessment terms, an EODM is likely to be SIL2.

An EODM is clearly fail-safe, as in the event of an obvious failure, the pilot would ignore the EODM and revert to the manual ODM.

MOD has recently procured a number of EODMs. In practice, it has proved very difficult to argue that an EODM program *on its own* has the required integrity, so effort was expended in showing that the cross-checks and surrounding processes make the use of an EODM acceptably safe. The problem of trusting an EODM in isolation can be summarised as *'even if one could gain sufficient confidence in the application code, the behaviour of the program would still depend on the correctness of the COTS operating system, for which that level of evidence is unavailable'*. This is not just about gaining access to the code, even if you did have access, the shear volume of code would make analysis for safety properties prohibitively difficult.

It should be noted that although the illustrated application is clearly a Microsoft Windows™ program, the same comments would apply to any windows-based platform.

2.2 Key properties of this case study

Whilst the following recommendations have come from analysis of this specific case study, they are presented in a generalised form so as to be relevant to an application with the following key properties:

- The application is fail-safe - an obvious failure of the application is not a hazard (though 'plausible but wrong' outputs may be);
- The user interface is essentially a fixed dialog - the same information always appears in the same places on the 'form';
- All relevant information is visible on the screen - i.e. the user can see the selected inputs and the outputs that correspond.

2.3 Recommendations from this case study

For some new application, even of a similar type, the safety analysis would need to be repeated. However, these recommendations should provide a guide to the sorts of issues that may arise and in many cases can be used as system requirements.

The recommendations are presented in roughly the order they arose in analysing the system (Pygott 2002). This tends to mean that the issues with the greatest impact on an application's development come late in the list. Conversely, some of the early recommendations are only really an issue in pathological cases or would only be an issue with unlikely design choices.

1. **Avoid use of primary colours for critical messages**: loss of an electron gun could cause undetectable loss of messages. This is especially significant if the display were *only* to use primary colours.

2. **Use colour graduation or a wedge shape to indicate that the VDU is scanning correctly, horizontally and vertically**: this is mainly an issue where the application displays regular rows or columns of data. Failure of the display scan logic might cause rows/columns of data to be exchanged or duplicated, masking critical data - improbable, but possible.

3. **All user input should cause an obvious change to the display**: basically, blind key entry should be avoided. In principle, someone looking over the

shoulder of the user should be able to justify the displayed outputs from the visible input data.

4. **Displayed outputs should only ever correspond to the displayed inputs:** any input changes should either cause the outputs to be recalculated, or if multiple inputs can be changed and then some 'calculate' behaviour triggered, the outputs should be removed as soon as one of the inputs is changed.

5. **An empty control should never be accepted as a default value:** this is significant for edit boxes, drop list, etc. The problem is that the control may be empty because the program has failed to refresh that part of the display (so is internally using a value different to the one the user thinks is implied by a default setting), or more generally the program and user may have different ideas of what the default is.

6. **A monitor program should be developed to run in parallel with the critical application:** see below.

7. **The execution of the monitor should be confirmed by a hardware watchdog:** again, this is expanded below.

2.4 Monitoring the application's outputs

Recommendations 1 to 5 above are essentially derived safety requirements for the application. However, even with all these in place, the application is still vulnerable to operating system errors: failure to perform some activity, error in library calculation etc. Most of these problems can be mitigated if the behaviour of the application could be independently monitored, and any errors reliably reported to the user.

The basic concept, therefore, is that a monitoring program acquires 'knowledge' of the inputs supplied by the user and the outputs produced by the application, and asks *'are the output produced consistent with the inputs supplied?'* A number of questions can immediately be posed:

*   how is the checking activity triggered?
*   how does the checker acquire the input/output data from the application?
*   how is an error (or success) reported to the user?
*   what prevents a latent failure of the checker masking failures of the application?
*   how can it be ensured that the checker calculates the input/output relationship in a way that avoids common cause failures with the application?

Note that the monitoring program is referred to throughout as the 'checker' rather than the 'monitor' to avoid confusion with the display, which also may be called the monitor.

How is checker activity triggered? There are two basic approaches that could be envisaged: autonomous and co-operative. For each, not only must the mechanism be considered, but also the effect of its failure, i.e. the checker triggers when it shouldn't, or fails to trigger when it should.

In a co-operative system, the application would trigger the checker by, for example, passing it a message after displaying the output. The main problem is that if this message is never sent or is 'lost by the operating system', the checker is never activated, and so no error can ever be reported. This is not necessarily a problem if, in addition to generating error 'messages' in some form, the checker also provides positive confirmation that the output is correct. However, the other disadvantage of this method is that it requires additional behaviour from the application, so cannot be used if the application is itself COTS.

In an autonomous system, the checker must be able to determine for itself, when it should start. For reasons to be explained in the next section, the proposed method to acquire the data from the application is for the checker to take a 'screen shot', i.e. to read the currently displayed bitmap from the video memory. This potentially provides a mechanism to trigger the checker activity. If the checker samples the screen at regular (and fast) intervals, it can look for the presence of information in the output positions as an indication that the application has generated some data.

How does the checker acquire the input/output data from the application? Again, the two basic approaches are autonomous and co-operative. The co-operative approach can be quickly discounted. Whereas with the triggering event, it may be possible to build an argument that says you can either trust the application to do it or detect if it fails to do so. Here there is the need to transmit complex data. The immediate question is what's to say that the data sent by the application to the checker is the same as it displays on the screen? This would seem to be an insuperable barrier to a co-operative approach.

The better alternative is for the checker to acquire the input/output data from the application independently. The recommended solution is for the checker to read the shared Video RAM memory and use Optical Character Recognition (OCR) techniques to rebuild the information seen by the user. Whilst this involves quite complex programming, it has the advantage that it is entirely independent of the generation of the information, so it is unlikely that complementary errors will occur. It also has the advantage that the path from the video RAM to the screen is independent of the information being displayed, so failures after the point where the values are checked are likely to be obvious.

How is an error (or success) reported to the user? The issue here is more on how to report success rather than failure. If the checker detects an error it can always (in theory) pop up a message box telling the user of the problem - with the caveat of considering the effect of the message not being displayed. Can the lack of such a message be taken as an indication that the checker accepts the result? Probably not, as it may just mean the checker is not running. A classic way of ensuring that a process is running is to use a watchdog. However, because the envisaged application is not running on a dedicated machine, the watchdog cannot be running all the time, but must be activated when the EODM starts and deactivated when it closes.

For reasons described in (Pygott 2002), a hardware watchdog can be envisaged that works as follows:

- At program initiation, the checker looks for the presence of the watchdog. It reports an error if it is not found. This is particularly important if it is attached to the processor as a peripheral (say via a serial or USB link) rather than being permanently installed;

- The checker performs any built-in tests (BIT) necessary to confirm that the watchdog is working, and reports an error if it isn't;

- The checker then activates the watchdog. Two enhancements to the activation process are:

  - activation should cause the watchdog's alarm to sound briefly (to demonstrate that the alarm is working). If this is deemed too intrusive, this feature could be made optional, but at the risk that an alarm failure may be latent and render the watchdog useless, though it may be that some initialisation BIT can go a long way to checking that the alarm is working;

  - activation could also return a random code value. This could then be used as a basis for the codes sent by the checker during the execution and termination phases. This reduces the probability of a program 'running wild', but looking to the watchdog like correct execution;

- During execution, the checker writes to the watchdog. Failure to write in a given period causes the alarm to sound. This means that if the checker detects an error, it can indicate this not only by presenting a warning message to the user, but also by stopping 'accessing the watchdog' as well (mitigating possible hazards with the operating system not displaying its warning message). If activation generates a code, this can be used as the basis for a specific value sequence that must be sent to the watchdog;

- At termination, the checker can 'turn the watchdog off'. Again, if a code was generated at activation, this can be used as the basis for a 'termination code', to prevent a wild running application accidentally turning off the watchdog.

What prevents a latent failure of the checker masking failures of the application or operating system? The main fault to be considered here is failure of the checker to run. If the checker is not expected to confirm the correctness of the output explicitly, then there is a potential problem of the checker being closed early or 'stalling'. In such circumstances, a mechanism, such as a watchdog timer, would be needed to confirm that the checker was active (the watchdog being enabled when the application is launched, and stopped when it closes). An alternative to a hardware watchdog would be for the checker to perform some continuous activity (such as playing an animation) to demonstrate that it is still active. This has a number of disadvantages over the hardware alternative, and so is not recommended:

- It relies on the user recognising that the checker stopping or not being displayed represents a safety hazard. Such negative behaviour from a program, other than the one that the user is concentrating on, is going to be too easy to ignore, especially when compared with an alarm sounding.

- The animation requires some screen 'real estate'. Whilst on most operating systems, a window can be made 'always on top' (i.e. always to appear above other windows) this is likely to be annoying to the user, and may still be obscured by modeless dialog boxes.

- If the animation is displayed using a library procedure, there is no guarantee that its continued operation demonstrates the continued operation of the monitor. It may well be launched as a separate thread, and its activity only confirms that this thread is active, not the monitor that launched it.

How can one avoid common cause failures with the application? Ultimately, when the checker has acquired the same input data as the application, it needs to confirm the outputs derived from that data. If it simply repeats the calculation performed by the application it will not detect systematic errors - though it would detect transient upsets.

The first consideration is the algorithm used to calculate the outputs. There are potential advantages in making the checking algorithm different to the one used in the application. It may be possible to use a far simpler algorithm to 'sanity check' the application's output, where the sanity check doesn't attempt to produce the same output, but can show that the application's output is 'near enough'. The main advantage comes if reducing the complexity of the checker's algorithm makes it more amenable to rigorous review or formal proof.

Even if a different algorithm is used, it may be still be difficult to avoid using common library functions. For example, if some output depends on the Log of wind speed, then it may be difficult to avoid calling the same library function with the same data, and being vulnerable to a data-dependent error. One method of avoiding this would be to modify the checking algorithm, by scaling or otherwise transforming the values used.

3 Case Study 2: an ATC case study

This case study is looking at programs that might be described as 'situation awareness' applications. The application takes information from some remote source and displays 'the situation' to the user.

The example used was constructed around an Air Traffic Control (ATC) demonstrator program, supplied by the Airspace Management Systems Department of QinetiQ, funded from the European Commission and illustrated in Figure 2 (see http://www.cordis.lu/telematics/tap-transport/research/projects/cincat.html). Again, this image is only illustrative, if the text is illegible it does not affect the argument.

The ATC application has a number of differences from the EODM application already discussed. The most significant is that the source of the information is a remote track processor that is accessed via some communications process. Whilst this aspect is discussed in (Pygott 2002), this paper will concentrate on the correctness of the display.

As far as the display is concerned, the most significant difference is that here information is provided by the position of icons on the screen, so the checker must in some sense 'look for' the information being displayed. Also it is quite legitimate for information to be masked, for example when aircraft paths cross at different flight levels. This means that the checker requires some form of memory and predictive behaviour.

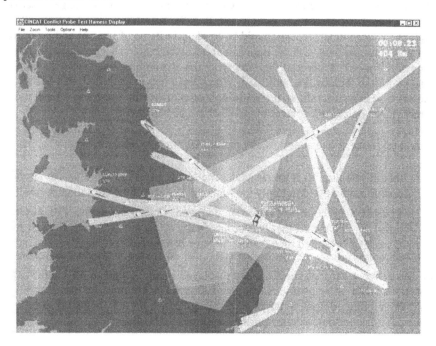

Figure 2: An illustrative ATC program

A study has been conducted into the techniques available to identify elements within the displayed scenes (Chisnall, Baker & Ducksbury 2002). A number of conclusions have been drawn from this study. For efficient image interpretation:

- items of interest should be in a limited set of fixed colours. This allows simple colour filtering to separate different types of displayed information.

- items of interest should be displayed by fixed icons or text from limited fonts. The size of the icons should be fixed even if the image is zoomed.

- the drawing order should be controlled so that the most significant items are always drawn over less significant ones, i.e. aircraft position - over labels - over fixed features - over the background. This removes the possibility of say an aircraft being obscured by a label, but does not prevent it being obscured by another aircraft.

- image identification can be efficiently performed by template matching, provided an initial search is performed to identify where targets might be.

It should be noted that the first two recommendations above are consistent with current ATC display design (*ODID IV Simulation Report* 1994).

The need for the monitoring process to perform a search to identify potential targets arises because template matching is inefficient when trying to identify a small image in a large area. So, if instead of searching the whole image, a set of potential target regions can be established, then the search time is a lot faster. The result of the searching process should be to identify items of interest and compare their positions and data with the information received by the checker from the track processor. For aircraft targets, this simply means ascertaining whether there should be an aircraft at the position indicated. Where label text is identified, this has to be associated with an aircraft, so the lead-line that connects an aircraft to a label must be recognised before the content of the label can be checked

In addition to the above activities, the monitoring process must check:

- that all the aircraft it knows about are being displayed (i.e. that the aircraft positions identified not only are accurate, but are all the aircraft expected);

- it must also cope with overlapping and obscured targets. For aircraft positions this means that loss or ambiguity of position must be allowed for, at least for a limited time, before an error is reported.

An alternative means of checking the displayed image would be to confirm the existence of data in the positions that are expected. For example, for each aircraft it could be confirmed that a target was being shown in the expected position. However, this does not address spurious additions to the display.

The recommended solution is therefore a hybrid one, which:

- uses colour filtering and a preliminary search to identify possible aircraft and label positions;

- uses template matching to confirm the identified data is correct;

- for any aircraft not so identified, looks at the expected position and sees if there is a target there (this copes with obscured and overlapping aircraft).

4 Case study 3: text and database processing

Aircraft operation and maintenance procedures are created as text documents. Such documents are in effect treated as attracting a significant SIL, with sophisticated audit and sign-off mechanisms being used to ensure that all units are working from the latest version of the printed document, and that all relevant personnel have been made aware of changes. With a printed document and physical signatures it is comparatively easy to be assured that the correct version of the document is being used, and that it has not been corrupted. If, for example, it was decided to have these procedures available electronically on a PDA, how could it be argued that the version seen was both up to date and correct?

For the GUI-based and situation awareness case studies, the strategy for ensuring the correctness of the displayed information was to provide a monitoring process to check what the user is shown. That approach will not entirely work for a database or text example, because only a fraction of the document is visible at any time (so for example the checker could confirm that some database record visible on the screen has been modified in accordance with the user's intentions, but not that some other records have not also been modified). So the fundamental question that arises is *when can the contents of the document be regarded as correct?* By analogy with critical paper documents, it is not usually sufficient for the author of the document to say that its correct, it has to be reviewed by some agreed process, and eventually annotated to show that it is an accepted version. Changes to the document will either go through the same review/approval process or a traceable change route from a previously approved version to the new version will be reviewed and approved. The document generation cycle can therefore be represented as shown in Figure 3.

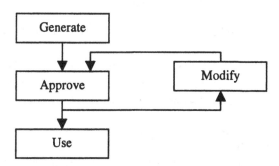

Figure 3: Document generation/review/use cycle

There is no reason why the process for electronically held critical documents should be any different. Indeed, as there are good reasons why the above process has come into being for paper documents (traceability, auditability, accountability etc.), one can argue that the process for electronic documents must be the same or at least provide the same benefits. From this point of view it makes sense to separate the generate-modify-approve cycle from use of an approved document.

Review and Approval Key to the approval process is how 'transparent' is the stored format of the document? If one considers a database or text document stored in a proprietary format, it is unlikely that any program other than the original editor can display it and there may be uncertainty as to the precise meaning of the information stored in the file (e.g. is there some control character embedded in the document that could cause the information to change?).

- **review using the same software that was used to create it**: this has to be the least satisfactory option, as it provides no protection against the majority of hazards. It is trusting the editing software to reliably display the data file. Whilst it may be argued that if the user is going to access the document in the same manner, if it appears correct to the reviewers, it ought to appear the same to the user, there is no guarantee that this is so because of the possible effect of embedded 'control characters'.

- **electronic review using different software**: this is a better option, as there is now an independence argument between the editing and review processes. There are potentially still problems if the data is recorded using a proprietary format: both in terms of getting an independent reader and knowing that the data does not contain embedded controls (unintentionally or intentionally) that could modify the data at a later read.

- **printing the data and reviewing the paper copy**: the key issue again is what software is used to print the data: the same as used to edit it, or different? If the data is only to be used in printed form (as in the aircraft maintenance example), it doesn't matter because the reviewed information is going to be the same as that used. However, if the intention is to use the information electronically, it is better that the printed copy is produced with independent software.

The two key recommendations from the above is that:

- critical data should be held in a non-proprietary and transparent form
- the data should be reviewed or printed for review using software independent from that used to create it.

Recording approval Having reviewed the document, it is important that the fact it has been authorised is recorded. By analogy with the paper process, this authorisation should be an integral part of the document, that makes the fact that an approved version is being used auditable. It is recommended that a digital signature is used to indicate authorisation (Diffie & Hellman 1976).

- To create a digital signature, some 'near-unique' property is extracted from the document, for example a 32-bit sum check over its contents, where there is a very low probability that a file with different contents will generate the same checksum. The authorisers then encrypt this checksum using a public key encryption scheme and embed the encrypted checksum in the document.

- An auditor or a piece of monitoring software can now confirm that the document has been approved, by using the public key to decrypt the checksum and confirm that the document's contents do indeed conform to that checksum. Because only the authorisers have the private key, the file cannot be undetectably modified (either deliberately or accidentally).

How do you know you are using the data correctly? Once a document has been approved and digitally signed to record the fact, correct use of the data is easier to establish. In effect the problem becomes similar to the EODM case study, in that a monitor program can open the document file and confirm the data presented on the screen agrees with that in the file. This may be either by direct comparison, if the application is simply presenting the stored data, or by inference, for example in where data from a database is used as input to a calculation.

If one were being particularly cautious, one may argue that opening two identical copies of the same file (or the same file twice for reading) is vulnerable to data-sensitive errors in the file access mechanism. A solution would be for the checker to access a transformed version of the database (e.g. numeric values scaled, boolean values inverted and character values encrypted with a substitution cipher). This still allows the checker to confirm the information presented to the user, but avoids any data sensitivities.

5 Conclusion

In all these case studies, hazard analysis has shown that safety-related applications can be justified on COTS platforms provided that:

- hazard analysis is used to identify the failure modes of the COTS that could lead to a hazard
- derived safety requirements are identified and implemented to limit the impact of errors in the application and platform
- monitoring software is used to 'sanity check' the application, to make all credible errors obvious and hence not a hazard
- the sound used on monitoring software implies the need for an independent hardware watchdog to ensure monitoring actually occurs.

The practicality of the monitoring software approach has been demonstrated by implementing the EODM example together with its monitor. On a 233MHz Pentium II under Windows NT™ the monitor has little visible impact on normal user actions (typing, mouse movement, etc). Both programs are available from the author.

6 Acknowledgement

The work described in this paper is funded under MoD Corporate Research Programme, "COTS software for High-Integrity Applications"

7 References

Bishop, P., Bloomfield, P. & Froome, P. 2001, *Justifying the use of Software Of Uncertain Pedigree (SOUP) in safety -related applications*, HSE contract research report 336/2001, London

Chisnall, J., Baker S. & Ducksbury, P. 2002, *Image Interpretation for Monitoring Safety-Related Software on a COTS Operating System*, Unpublished QinetiQ report: QinetiQ/KI/SEB/WP020349/1.0, Malvern

Defence Standard 00-56, 1996, *Safety Management Requirements for Defence Systems*, Issue 2, Ministry of Defence, Directorate of Standardization, Glasgow

Defence Standard 00-58, 2000, *HAZOP Studies on Systems Containing Programmable Electronics*, Issue 2, Ministry of Defence, Directorate of Standardization, Glasgow

Diffie,W. & Hellman, M. 1976, 'New directions in cryptography', *IEEE Transactions on Information Theory*, vol IT-22, No 6, pp 644-654

Jones, C., Bloomfield, P., Froome P. & Bishop P. 2001, *Methods for assessing the safety integrity of safety-related Software Of Uncertain Pedigree (SOUP)*, HSE contract research report 337/2001, London

McDermid, J. 1998, 'The Cost of COTS', *IEEE Computing*, pp46-52

ODID IV Simulation Report 1994, EEC Report 269/94, Eurocontrol Experimental Centre, Bretigny sur Orge

Pygott, C. 2002: *Case studies for the development of safety related applications on COTS platforms*, Unpublished QinetiQ report QINETIQ/KI/SEB/CR020311/1.0, Malvern

Scott, J., Preckshot, G. & Gallagher, J. 1995, *Using Commercial-Off-The-Shelf (COTS) Software in High-Consequence Safety Systems*, Lawrence Livermore National Laboratory, Fission Energy and System Safety Program Report: UCRL-JC-122246, Livermore

APPLICATION AND DEVELOPMENT
OF STANDARDS

The Application of BS IEC 61508 to Legacy Programmable Electronic System

H M Strong CEng MIEE MInstMC
D C Atkinson MIEE

1   Introduction

British Nuclear Fuels plc (BNFL) is an international nuclear energy business serving Governments and nuclear utilities worldwide. Combining international reach with world-class technology and skills, it's operations span the entire nuclear energy cycle from fuel manufacture, reactor design and services, electricity generation, waste management through to decommissioning and environmental services.

The Sellafield Site in West Cumbria has been at the centre of the UK's civil nuclear programme since its inception in the mid-1950's under the auspices of the United Kingdom Atomic Energy Authority (UKAEA). BNFL took over the operation of the site upon its formation in 1971.

With a continuous focus on safety for its people and the environment, BNFL encourages the development of systems designed to maximise remote working, and minimise radiation doses to its employees. It is no surprise, therefore, that the company has for several decades developed and maintained innovative and functional Programmable Electronic Systems (PES) to monitor, manipulate and control all areas of its process plant. Figure 1 illustrates the complete spectrum of PES that are employed on the site ranging from specialist radiological instrumentation through to Management Information Systems. These systems reside on a wide range of platforms using many different computer languages and configuration tools. The Sellafield site has successfully developed and maintained a Plant Safety Case for each of its operational plants, and has PES identified to play varying roles within these Safety Cases.

Following the emergence of new international standards impacting PES Safety Systems, BNFL carried out a thorough review of the procedures controlling software modifications to the PES, the people modifying the PES and the function of the PES in relation to the Safety Case. The review resulted in a project to upgrade the procedures and working practices used on all PES to ensure BNFL designs and maintains PESs in line with modern standards.

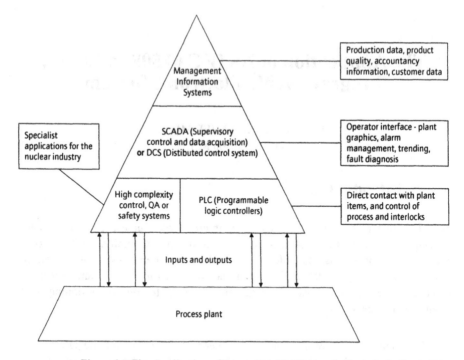

Figure 1 – The Application of Programmable Electronic Systems in Control Systems

2   Project Scope

Figure 2 diagrammatically shows the final project scope with three distinct workstreams, which are described in Sections 2.1 to 2.3.

Numerous national and international standards were reviewed during the project with the developed procedures and processes being ultimately aligned with:

BS IEC 61508 [1]   Functional safety of electrical/electronic/programmable electronic safety–related systems.

BS ISO/IEC 12207 [2]   Information technology - Software lifecycle processes.

BS EN ISO 9000-3 [3]   Quality management and quality assurance standards – Part 3. Guidelines for the application of ISO 9001:1994 to the development, supply, installation and maintenance of computer software.

HSG65 [4]   Successful health and safety management.

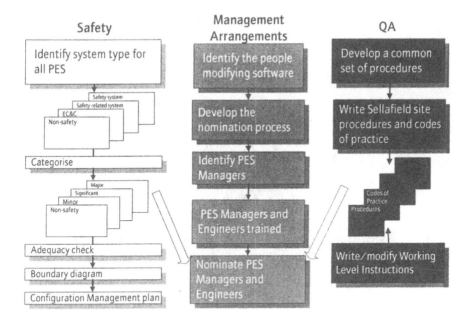

Figure 2 – PES Improvement Workstreams

Key learning – There is a large degree of overlap between the requirements of BS IEC 61508 and the quality standards. Careful alignment of practices must ensure that duplication of work does not occur.

The adoption of new procedures and working practices in any industry involves cultural change among the impacted staff. This project was no different and was run as a major change programme. Figure 3 illustrates the overlap and intersection of the relevant standards.

2.1 Quality Assurance (QA)

2.1.1 Goal

To write a single set of procedures that align with the standards indicated for the design and modification of all safety and safety related PES and to baseline the systems, from which point the new arrangements can be applied.

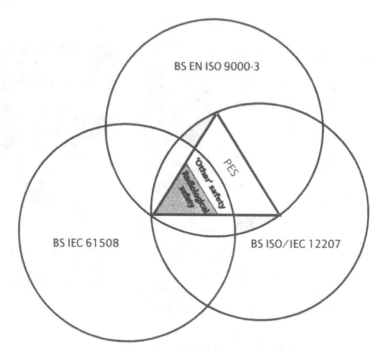

Figure 3 - The Application of Relevant Standards on PES

2.1.2 Challenges

The procedures must:

- Be applicable to both new and legacy systems;
- Allow for a common and consistent process to be deployed across the Sellafield site;
- Be applicable to all platforms and languages;
- Allow the removal of all existing local procedures and working practices; and
- Align with the primary processes which maintain safety on the site, such as, Plant Modification Proposals (PMP) and Safe System of Work processes;

Key learning – Although the application and type of PES can vary considerably, for example:the degree of safety significance; the hardware platform and the programming language; one modification process has been developed which applies the same principles to all.

2.2 Safety

2.2.1 Goal

To identify every PES employed on the Sellafield Site, and categorise them such that the newly developed procedures can be properly applied based upon their importance or required integrity.

This is straightforward for new systems as BNFL will design and build all new PES based upon the Safety Integrity Levels (SIL) required for the PES. For legacy systems this requires assessment of all the existing systems to allocate an appropriate SIL equivalent designation. Safety Integrity Levels target failure measures are shown in Table 1.

Safety Integrity Level	Target failure measures for an E/E/PE safety-related system	
	Low demand mode of operation average probability of dangerous failure on demand	High demand/continuous mode of operation probability of dangerous failure per hour
4	$>=10^{-5}$ to $<10^{-4}$	$>=10^{-9}$ to $<10^{-8}$
3	$>=10^{-4}$ to $<10^{-3}$	$>=10^{-8}$ to $<10^{-7}$
2	$>=10^{-3}$ to $<10^{-2}$	$>=10^{-7}$ to $<10^{-6}$
1	$>=10^{-2}$ to $<10^{-1}$	$>=10^{-6}$ to $<10^{-5}$

Table 1 – Safety Integrity Levels–Target Failure Measures for a Safety Function
(Source: Ref. [1])

2.2.2 Challenges

To produce a system of categorisation for PES that allows the appropriate application of procedures to both new and legacy systems.

To review the existing Plant Safety Cases to ensure that the safety function of each PES is clearly identified, and categorised to the system above.

To carry out a thorough review to confirm that the PES is adequate to perform its declared safety function, and to produce boundary diagrams and configuration management plans to record the scope and extent of each PES.

Note: The site has 72 plant safety cases and 1237 PES representing a huge task. To give an appreciation of scale, one system out of the 1237 PES is identified as one of the largest Distributed Control Systems (DCS) in Europe.

2.3 Management Arrangements

2.3.1 Goal

To identify all the staff involved in the PES design and modification process, train them in the use of the procedural suite and formally assess and nominate them against their duties and responsibilities.

Note: Nomination and appointment of the personnel was aligned to the BNFL process for appointment of key technical and safety responsibilities. (Reference was made to the BCS/IEE competency framework although it was not followed in full).

2.3.2 Challenges

Any change project involving several hundred staff needs a programme which allows time to communicate the changes and to obtain buy in.

Key learning – Writing procedures to BS IEC 61508 appears to be the primary activity on a project of this nature. However, the true balance of workload is 'write the procedures – 25%', 'categorise and adequacy check the PES – 20%', 'train and ingrain the arrangements in the staff and ensure competence – 55%'.

3 PES System Type and Categorisation

A major challenge was to develop a process that would allow legacy and new PES to be managed and modified by the same set of procedures. A review of national standards, international standards and other nuclear operators definitions of safety systems failed to identify common terminology. For example, the use of the term 'safety- related' has differing meanings in BS IEC61508 and IAEA standards.

The definitions for our systems were agreed as those below. Whilst naturally biased towards radiological safety, we did allow for a PES type that had environmental, conventional and chemotoxic (E,C&C) safety functions within it.

Safety System
'A system which acts in response to a fault to prevent or mitigate a radiological consequence'.

Safety-related System

'A plant system, other than a safety system, on which radiological safety may depend'.

Non-safety

'A plant system on which there is no dependence for radiological safety'.

Environmental, conventional, chemotoxic (E, C&C)

'A system that provides a sizeable contribution towards environmental, conventional or chemotoxic safety protection'.

This range of definitions allowed us to clearly define the type of function the PES carries out and manage it accordingly. The differentiation between types allowed BNFL to recognise PESs which initiate fault sequences as well as those PES being used for protection or mitigation and, therefore, manage them appropriately.

Having identified the system type it was also necessary to identify the level of reliability, dependence or integrity required of the PES.

The structured process required to apply BS IEC 61508 is based upon SIL. Given that legacy systems have not been developed to a particular SIL standard then a system of categorisation was required which allowed an equivalent to the SIL identification to be assessed and allocated to each PES.

Figure 4 shows the basis of the system that allows both old and new PES to be managed by common procedures.

The key component of this approach was the use of a Procedural Rigour Level which increments from 'Non-Safety', to 'Minor' (SIL0 area and SIL1), to 'Significant' (SIL2) and finally to 'Major' (SIL3 and SIL4). These Procedural Rigour steps or increments were designed to be aligned with important breakpoints in the BS IEC 61508 standard.

Systems were then categorised according to a set of criteria shown in the flowchart in Figure 5. This method caters for PES where the Plant Safety Case declared a numeric safety dependence or where a qualitative claim is made. This dual approach was essential to deal with the differing safety case methodologies applied historically.

No dependency claim in the safety case	Low dependency claim in the safety case		SIL 1		SIL 2		SIL 3	SIL 4
	Low demand mode	$1.0E^{1}$		$1.0E^{2}$		$1.0E^{3}$	$1.0E^{4}$	$1.0E^{5}$
	High demand mode	$1.0E^{5}$		$1.0E^{6}$		$1.0E^{7}$	$1.0E^{8}$	$1.0E^{9}$

New systems

Non-safety	SIL 0		SIL 2	SIL 3	SIL 4

Existing/legacy systems

Non-safety	Minor	Significant	Major

Procedural rigour

Non-safety	Minor	Significant	Major

Figure 4 - The categorisation diagram for both new and legacy PES

Note: SIL 0 does not exist as a specific category, its use was introduced to align with low dependency system.

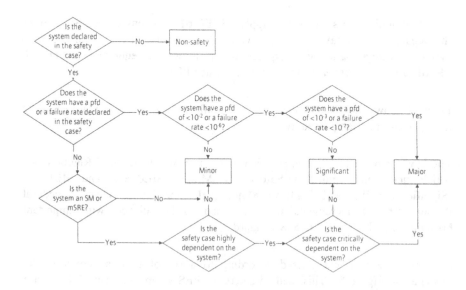

Figure 5 – Categorisation Assessment Flowchart for Legacy PES

Note: SM – Safety Mechanism and mSRE – minimum safety-related equipment. These are generic categories of safety system used within the Plant Safety Cases.

Key learning – The adequacy confirmation process checked the validity of each individual use of the PES in the safety case, reviewed any modifications made for possible adverse impact on the safety function and confirmed the PESs ability to deliver its safety function.

The resulting review of Plant Safety Cases then:
- Identified the safety function of each PES;
- Gave the PES a Safety System Type; and
- Categorised the PES with a non-safety, minor, significant or major Procedural Rigour Level.

Along with other key safety information for each PES this data was entered onto a register. The register data is populated by designers and safety case experts, and is accessed and read by modifiers and operators.

Certain principles were then included in our design methodology:
- If a claim is made for a specific SIL or category for any part of the PES, then the whole PES must be designed, tested and maintained at that level;
- Ensure there are no protection circuits in the basic plant control system;
- Claims for basic plant control systems will be restricted to 'SIL 0' area;
- The use of PES in the Safety Case will be minimised, for example, by 'significant' PES being driven down to 'minor' or 'non-safety' by making claim against other protection devices in the fault tree; and
- The safety margin will always be maintained.

Key learning - Each individual safety function within a PES is clearly understood by Plant Safety Case owners, plant owners and PES modifiers alike. With the known benefits of placing a PES at a lower Procedural Rigour Level all parties now work together to encourage the use of other means of risk reduction using e/e or other technology, for example, mechanical interlock.

There are substantial costs associated with the development and maintenance of high reliability software systems. Therefore, the aim is to reduce the level of reliance placed upon these systems to minimise the lifetime costs.

4   Adoption of BS IEC 61508

The primary technical standard considered in the development of the arrangements for the modification and management of PES was BS IEC 61508. This standard provides a generic safety lifecycle which encompasses the design and realisation of PES based upon the use of a structured process and of appropriate methods and techniques. Generally these are required to be selected and incorporated into a safety plan on a project by project basis. This is done using a risk based approach where the robustness of the methods, techniques and levels of independence increase with the required SIL to minimise the introduction of systematic faults and maximise their capture. This standard is based primarily around the design of new systems and the level of guidance on how it should be applied to modifications is minimal, and practically non-existent for application to legacy systems.

For modifications to a PES to be fully compliant the standard would require arrangements to have been in place from conception, for example, so that design documentation is in place to the required standard. The application to legacy systems has required pragmatism to ensure the processes and methods used are as closely aligned to the standard as practicable, but deal with systems developed before the standard was introduced. The process BNFL have developed allows for full compliance for new systems and alignment for it's legacy systems.

Whilst the arrangements necessary to apply BS IEC 61508 are targeted at safety PES, the main elements of the lifecycle proposed are also covered by good quality arrangements advocated by other standards such as BS EN ISO 9000 and BS ISO/IEC 12207. Therefore, the procedures have been developed such that the arrangements also address these quality requirements and form a single set of working processes to be applied to modifications to all PES.

The Procedural Rigour Level allocated against a PES can then be used through all the arrangements for PES management, from change control to the competency and nomination of staff.

Increasingly robust methods are applied as the Procedural Rigour Level of the PES increases. Whilst the Procedural Rigour Level has been developed to align SIL requirements to legacy systems, the allocation can also be applied to other PES with important functions. This means appropriate levels of Procedural Rigour can be applied to systems based upon their level of business risk for example, product quality recording of manufactured product, environmental protection or nuclear materials accountancy. The important function being dealt with in the same way as safety functions are within BS IEC 61508.

BS IEC 61508 requires the standard to be considered for each project. The lifecycle, methods and techniques to be used are recorded in a safety plan which is then adhered to for the project. This methodology is impractical to apply to the management and modification of over 1200 PES by 300 engineers on a modification by modification basis. To do this would require each engineer to have a detailed working knowledge of the standard and put forward a safety plan for each change. This would lead to diverse modification lifecycles and methods, techniques and tools being used across all the systems. To avoid this a standard process is required using a standard set of methods and techniques. This also enables engineers to be deployed across a large number of systems and the strategic rationalisation of platforms, methods and tools in use.

Key learning – Corporate review and selection of the BS IEC 61508 tables allows consistent application of the techniques selection, and gives efficiencies in the training of staff in these techniques. This generally removes interpretation of this complex standard by an individual engineer.

A generic process was developed which incorporates the modification requirements and integrates them into the modification/realisation lifecycle detailed in BS IEC 61508. The appropriate methods and techniques to be used in this lifecycle have been selected from the tables in the standard and incorporated into the process. In this way the PES engineers can follow a generic process using a rationalised set of methods and techniques for all modifications. This eliminates the need for engineers to consider the detail in the standard for each modification and the potential divergence of process and methods.

Wherever possible proforma and checklists were developed to aid the engineers by deploying a common process and assist with its consistent application. For example, detailed forms have been developed which guide engineers through the Design Reviews, Impact Analysis and Functional Safety Assessments necessary to support each modification, as well as general process forms to ensure the lifecycle is followed and all work is suitably documented. The procedures, form and checklists become the generic documented modification safety plan required by the standard and negates the need to develop one for each modification. The process developed delivers a SIL equivalent process for all modifications.

An increasing amount of detail, robustness of methods and level of independence is required as the Procedural Rigour Level of the system increases. These increases are aligned to the requirements to support the increasing dependency on the system and SIL. Each method and technique in every table in BS IEC 61508 has been considered on its merits and effectiveness and an appropriate (not the minimum) set selected. This provides a prescriptive set of methods for which each engineer becomes proficient in the application of. An example of the selection of methods and techniques is shown in Tables 2 and 3.

Key learning – A number of the methods proposed by BS IEC 61508 are not widely used or are open to inconsistent application. Gathering the correct personnel and specifying usage is essential to selecting an appropriate set of techniques.

Technique/Measure	SIL1	SIL2	SIL3	SIL4
1 Boundary value Analysis	R	R	HR	HR
2 Checklists	R	R	R	R
3 Control flow analysis	R	HR	HR	HR
4 Data flow analysis	R	HR	HR	HR
5 Error guessing	R	R	R	R
6 Fagan inspections	---	R	R	HR
7 Sneak circuit analysis	---	---	R	R
8 Symbolic execution	R	R	HR	HR
9 Walk-through/design reviews	HR	HR	HR	HR

Table 2 -The Static Analysis requirement (Table B8) of BS IEC 61508 (Source: Ref. [1])

Procedural Rigour Level	Non-safety	Minor	Significant	Major
Static analysis techniques	• Checklists • Walk-throughs/ design reviews	• Checklists • Walk-throughs/ design reviews	• Checklists • Walk-throughs/ design reviews • Boundary value analysis • Control flow analysis • Data flow analysis	• Checklists • Walk-throughs/ design reviews • Boundary value analysis • Control flow analysis • Data flow analysis

Table 3 – Static Analyses Requirement versus Procedural Rigour Level

However, for legacy systems the use of a specific method or technique may not always be possible or applicable. In this case justification for the use of an alternative method is developed as part of the modification and assessed as part of the Functional Safety Assessment (FSA).

Key learning – Where a prescribed method is not used this must be justified and alternatives considered. This is reviewed as part of the FSA to ensure the robustness of the process has not been degraded.

Another key area where legacy systems are likely to require additional guidance is in the support information underpinning the modification process particularly in the availability and quality of design documentation. Processes were established and implemented to ensure adequate safety and design information is in place for all systems; this has been crucial to the ability to manage the PES appropriately.

Further detail on the implementation of these principles is given in the remainder of the document. The primary focus of the information presented is on the modification process itself and is described in the following sections.

5   The Modification Process

Within the overall BS IEC 61508 safety lifecycle, modification to PES software is described within the Overall Modification and Retrofit Phase (phase 15). The main elements of this phase are to:

- Ensure a modification is initiated by an authorised request;
- Carry out an Impact Analysis, which includes an assessment of the affect of the modification on the safety function;
- Identify the appropriate overall safety lifecycle phase that the modification should return to;
- Document the Impact Analysis and all the details of the modification; and
- Authorise the realisation of the modification to commence.

Figure 6 shows a model on which to build a modification proposal process.

The modification assessment process considers the appropriate phases within the overall safety lifecycle that must be returned to. As a minimum the process must ensure that the Software Safety Realisation phase is carried out. The key elements of the Realisation phase are detailed in BS IEC 61508-3 Software Safety Lifecycle requirements, and is shown as an overview in Figure 7.

The generic process developed is based upon the Modification Procedure model within BS IEC 61508-1 and the Software Safety Realisation lifecycle within BS IEC 61508-3. The two phases have been amalgamated to form a single modification/realisation process as shown in Figure 8.

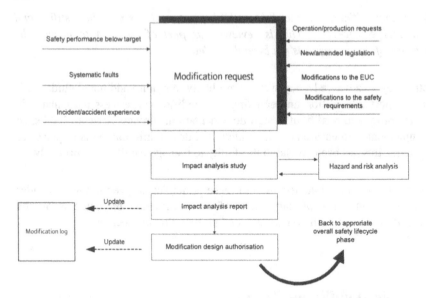

Figure 6 – Example Modification Procedure Model (Source: Ref. [1])

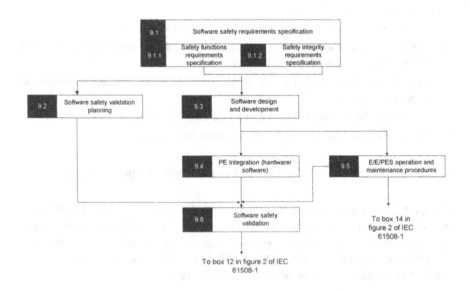

Figure 7 - Software Safety Lifecycle (in Realisation Phase) (Source: Ref. [1])

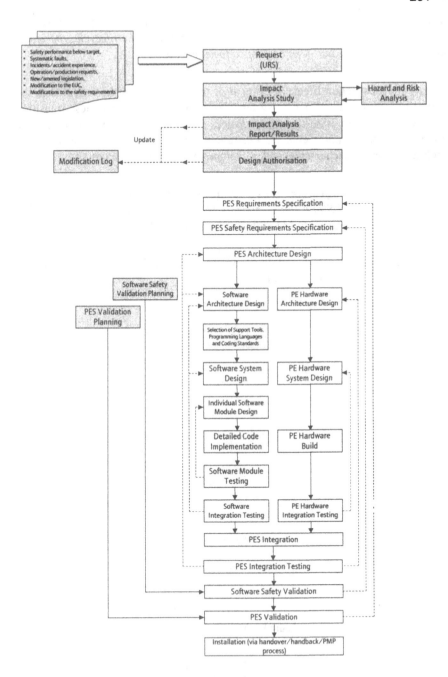

Figure 8 – Combined Modification/Realisation Lifecycle

232

Key learning – The majority of modifications are reasonably straightforward and simple, usually only returning to the realisation phase of the lifecycle, therefore, the standard process has been developed to reflect this.

Clearly the Realisation lifecycle element of this process must be reflected in the company arrangements for new system design to facilitate a seamless transition between system provision and maintenance. This allows for new systems to be aligned into the requirements for Operations and Maintenance (using common tools and support mechanisms) and, therefore, minimise lifetime costs.

Key learning – The lifecycle, tools and methods used in the initial design and realisation of the PES and the documents produced must align to those used in Operations, Maintenance and Modification to minimise lifetime costs.

A key element is the design and development model used in the process, which is a modified 'V'-model', as shown in Figure 9.

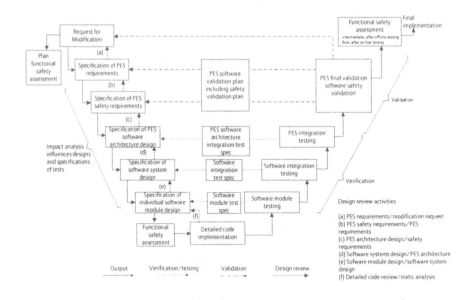

Figure 9 – Modified V-Model

The only changes made to the V-Model and the Software Safety Realisation Lifecycle proposed by BS IEC 61508-3 are purely to add standard user requirement specifications to those for safety and to include details of the verification and validation documentation required.

The process developed also ensures that the modification integrates into the existing plant safety procedures, particularly when installing and validating the modification to the PES on plant. As the on-plant testing is designed well before it is required, the impacts upon the plant are fully understood by the plant owners and appropriate plans can be implemented.

The key elements of the modification/design procedures are:
- Clear knowledge and documentation of safety functions of the Plant and PES;
- Communication of safety and technical issues to all parties;
- Competency of personnel;
- Structured software lifecycle using the V-model for design and testing;
- Impact Analysis/risk assessment of modifications;
- Functional Safety Assessment process;
- Production and maintenance of appropriate detailed documents for the PES;
- Appropriate levels of independence throughout the process;
- Detailed guidance to modifiers on the methods, tools and techniques;
- Use of Procedural Rigour Levels to align new and legacy safety systems and other systems with other important functions to a SIL equivalent process; and
- Assessment and control of contractors used for modifications.

The procedural modification process derived from these requirements is divided into three main elements:
- System Modification Assessment (SMA) – initiation, impact assessment and authorisation of modifications;
- System Modification Record (SMR) – detailed requirements analysis, design, test production and coding; and
- System Release Note (SRN) – installation, integration with plant PES Safe Systems of Work and final functional and safety validation.

Supporting these three main elements are six support processes:
- Impact analysis;
- Verification and validation;
- Design review;
- Functional Safety Assessment;
- Configuration management; and
- Documentation.

Each element is described in more detail in sections 5.1 to 5.4.

These processes are all proforma and checklist driven to provide a consistent framework for the engineers and to minimise the need for reference to the standards.

The procedural suite for modification is shown in Figure 10, which also shows the integration to the site plant safety arrangements. The overview of how all these elements are co-ordinated together is shown in Figure 11.

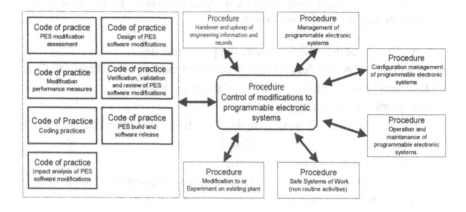

Figure 10 - The Modification Procedural Suite

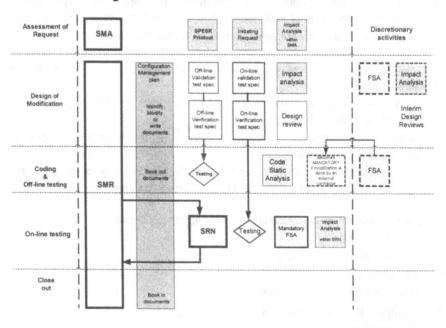

Figure 11 - Modification Process Overview

Note: SPESR is the Sellafield Programmable Electronic System Register which contains key details of all PES on the Sellafield Site.

Key learning - Alignment with the existing Safety Case/HAZOP/HAZAN processes and those for handover and handback of plant systems is essential and has been integrated into the whole process to minimise duplication

Key learning – The adoption of the basic principles by our software contractors has allowed for very clear specifications to be agreed between both parties if a modification, design or build is outsourced. However regular review by BNFL staff ensures that the ability of the PES to perform its safety function is not threatened during the modification process.

5.1 System Modification Assessment (SMA)

The purpose of the System Modification Assessment (SMA) is to address the requirements of the Modification Process Model outlined in BS IEC 61508-1 and Figure 6 of this paper. The SMA is based on a single form which is used to:

- Capture the details of the modification requested and details of the target PES;
- Consider the validity of the request and identify any other possible solutions;
- Carry out an impact analysis, based on a series of checklist questions, to identify the scope of the modification, any other systems potentially affected and most, importantly any potential impacts upon the safety function of the target PES;
- Specify the safety validation required to ensure the safety function is unaffected by the modification. For example, any known functional tests; and
- Estimate the personnel resources required, approximate timescales and costs to carry out the change.

Where potential to degrade a safety margin (see section 5.4.1) is identified in the Impact Analysis section a detailed safety assessment process is initiated to consider the modification and its effects upon the Plant Safety Case and existing HAZOP/HAZANs.

The SMA is conducted in partnership by the PES Engineer/Manager and the plant personnel. This ensures that appropriate technical and plant safety knowledge is used to assess and authorise the modification.

Subject to satisfactory conclusion of these considerations the modification is authorised to proceed to design and development.

Key learning – Successful deployment of the SMA process gives a far more challenging review of the safety implications and necessity of any modifications, at an early stage in the modification process, and has resulted in a 50% reduction in the overall number of modifications made.

5.2 System Modification Record

The purpose of the System Modification Record (SMR) is to address the requirements of the Software Realisation Lifecycle outlined in BS IEC 61508-3 and shown in Figure 7 of this report. The process to be carried out is shown in Figures 8 and 9. The SMR is based upon a single main form with several modular sub-forms which are included as required, for example, Design Review, Verification and Validation Plan, Functional Safety Assessment (FSA).

The SMR process ensures that:
- The requirements for the modification are correctly specified and then cascaded into the appropriate design documents and considered in a structured way through the 'V'model;
- A detailed impact analysis is carried out to: identify the scope of the detailed modification throughout the system; which modules of code are affected; identify any defensive programming required; and, the scope of verification and validation to be applied. This may be revisited at each phase of the 'V' model;
- Test plans and schedules are developed concurrently with each design phase (and not retrospectively);
- A Design Review is carried out for all phases of design. These may be carried out after each phase or rolled up into a single design review prior to code implementation dependant upon the complexity and category of the change;
- Code assessment and static analysis is carried out to confirm that the implementation addresses the design and conforms to the appropriate coding standards; and
- Functional Safety Assessments are carried out to confirm the safety margin of the system has not been degraded by the modification.

The SMR acts as the primary controlling document for the technical modification to the PES and interfaces with the PMP (plant modification procedure) process where potential safety impacts have been identified or the modification has multi-disciplinary considerations.

Once all these activities and any off-line verification and validation required is complete then the modification can be authorised for integration upon the target system. This is managed and controlled by the System Release Note (SRN) described in the following section.

However, the SMR is not closed out until the SRN is completed and all documentation is formally updated.

5.3 System Release Note

The purpose of the System Release Note (SRN) is to co-ordinate the installation and testing of the modification on the target plant PES, interfacing with the appropriate plant safety management processes as required. The SRN is based upon a single form which is used to:

- Ensure that a pre-implementation impact analysis is performed during the system build and release process. To consider any risks associated with the implementation, combination of concurrent modifications or plant status, and to ensure that appropriate controls are applied;
- Define the interface with the plant safe system of work procedures;
- Ensure the verification and validation plans are carried out and reviewed;
- Ensure that Functional Safety Assessment is performed prior to plant restart;
- Define the point and criteria at which hand back/plant restart may occur;
- Co-ordinate multiple SMRs into a single release to plant;
- Ensure that there is an assessment of the cumulative impacts of multiple SMRs released to plant at the same time; and
- Ensure contingency plans are in place to recover the PES to a known position should validation be unsuccessful.

The SRN records the outcome of all verification and validation and ensures that the results are considered within a final FSA to confirm the plant is safe to restart.

5.4 Support Processes

A number of sub-processes are required to support the modification process. Generally, they are modular processes which are used as required within the lifecycle. Details of these processes are described in the following sections.

5.4.1 Impact Analysis and Safety Margin

The key elements of the impact analysis processes are to:

- Consider the effects of the modification on the functional safety of the PES;
- Provide an indication of the impacts produced by a modification;
- Guide design and scope of testing to ensure impacts and risks identified are defended against and validated as not occurring; and
- Consider whether a safety margin has been reduced and if so then to ensure a hazard and risk analysis is performed via the plant modification proposal (PMP) process.

Three impact analysis are carried out:

- Initial during SMA;
- Detailed during SMR; and
- Pre-implementation during system build and release (SRN).

Three separate checklists are used to ensure consideration is given to the differing potential risks and impacts present at each of the three phases.

This process ensures that any potential reduction in the safety margin provided by the PES is appropriately considered and justified in terms of ALARP (As Low As Reasonably Practicable). A description of safety margin is given below and shown in Figure 12.

When a hazardous event is identified, the product of the consequences and frequency of the event provides the *initial risk*. This initial risk may be higher than that which is considered to be *tolerable* by society and current legislation. In order to reduce the risk to a tolerable level, PESs may be employed as part of a suite of risk reduction measures.

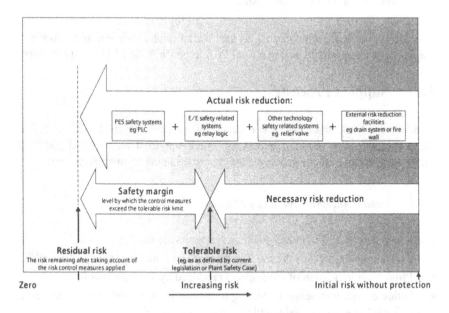

Figure 12 – An illustration of Safety Margin

The risk which remains following deployment of these measures is known as 'residual risk', with the extent to which the measures exceed the tolerable risk being known as the 'safety margin' and must be appropriately considered.

All PESs identified and categorised as safety systems provide part of the actual risk reduction. Therefore, any modification to a safety system could change a safety margin.

The incorporation of a hazard and risk analysis study into the software modification lifecycle is covered by the Impact Analysis and the PMP process. The initial key phrase questions in Impact Analysis assess whether the modification reduces or may reduce a safety margin. If this is the case then a PMP must be raised and a safety assessment is carried out to specifically consider the impacts. The safety assessment determines to what extent the margin is reduced and presents and justifies the findings in terms of the overall risks being ALARP. The impact analysis takes place before any changes to the plant system are allowed to be performed.

5.4.2 Verification and Validation

The production of Verification and Validation (V&V) plans and schedules is ensured within the SMR process and the use of them in the SRN. The key elements of V&V are to:

- Ensure a plan for verification testing is compiled concurrently with the development of the phase;
- Ensure documentation of the criteria, techniques and tools to be used in the verification activities;
- Ensure that verification testing is carried out according to the verification plan.
- Define how to produce a validation plan and tests;
- Ensure software safety validation and PES final validation activities are carried out in accordance with the overall validation plan for the PES system; and
- Ensure system details and test results are recorded, and validation reports produced, in accordance with the overall validation plan.

The methods and techniques used to address V&V have been derived from tables A5 to A9 in BS IEC 61508-3 and these are prescribed to the engineer and included within the procedures. An example is illustrated in Table 4.

Procedural rigour level	Non-Safety	Minor	Significant	Major
Software validation and software safety validation testing	Functional and black box testing using: • Boundary value analysis • Equivalence classes and input partition testing	Functional and black box testing using: • Boundary value analysis • Equivalence classes and input partition testing	Functional and black box testing using: • Boundary value analysis • Equivalence classes and input partition testing Probabilistic testing must be applied at the validation stages, using automated test tools where available.	Functional and black box testing using: • Boundary value analysis • Equivalence classes and input partition testing Probabilistic testing must be applied at the validation stages, using automated test tools where available

Table 4 – Software Validation Requirements versus Procedural Rigour

5.4.3   Design Review

Each modification is subjected to a design review to:

- Assess the design work for its technical quality;
- Confirm that the design satisfies the User Requirement Specification (URS) and Safety Requirement Specification (SRS);
- Confirm that safety functions are not being affected unintentionally; and
- Confirm that the relevant coding standards and styles have been followed.

The attendance of the Design Review Panel is based upon the complexity of the modifications and Procedural Rigour Level of the system and is illustrated in Table 5.

	Non-safety/Minor	Significant/Major
Low complexity modification	• Independent PES Engineer peer review.	Review panel consisting of: • Independent PES Engineer. • The PES Manager. • The Design Authority. • Technical expert (as required). • PES Managers of all affected systems (as required).
Medium complexity modification	Review panel consisting of: • Independent PES Engineer. • The PES Manager.	Review panel consisting of: • Independent PES Engineer. • The PES Manager. • The Design Authority. • Technical expert (as required). • PES Managers or nominees of all affected systems (as required).
High complexity modification	Review panel consisting of: • Independent PES Engineer. • The PES Manager. • PES Managers of all affected systems.	Review panel consisting of: • Independent PES Engineer • The PES Manager. • The Design Authority. • Technical expert (as required). • PES Managers of all affected systems (as required).

Table 5 – Design Review Panel Membership Requirements versus Procedural Rigour

The design review process is based upon a checklist to ensure consistency. It can, either, be carried out following each design phase or 'rolled up' with all phases being reviewed prior to coding. It must be recognised that there is an element of risk associated with carrying out a 'rolled up' review. Primarily in the amount of potential rework required following an unfavourable outcome, but also with the potential to miss errors due the amount and complexity of work being assessed. For low complexity changes the risk is generally low and can remove several delay points from the process.

5.4.4   Functional Safety Assessment (FSA)

The purpose of the Functional Safety Assessment (FSA) is to consider the process followed, the methods and techniques used, the competency of the personnel and the outcomes of the modification to make a judgement on the PES continuing to fulfil its safety dependency (the modification doesn't reduce the safety margin).

Again a checklist form is used to ensure all the appropriate inputs are considered and used to make the judgement.

The independence of the person leading the assessment is based on Procedural Rigour of the PES and illustrated in Table 6.

	Non-safety	Minor	Significant	Major
Level of independence of FSA panel leader	Not required	Independent person	Independent department	Independent company

Table 6 – FSA Panel Membership Requirement versus Procedural Rigour

The FSA panel leader must ensure appropriate representation is obtained to allow the FSA to be conducted. Up to three FSAs can be carried out through the process or the whole process can be 'rolled up' prior to reintroducing the hazards to the plants.

5.4.5   Configuration Management

The purpose of the configuration management process is to:

- Facilitate identification of a complete, current PES baseline so that everyone working on a PES at any time in its lifecycle uses correct and accurate information;
- Ensure changes are managed and controlled; and
- Provide the ability to return to a previously validated baseline.

To support this a Configuration Management Plan (CMP) has been produced for every PES which details all the elements which comprise the PES (hardware, software and documentation). This document provides key evidence and allows for detailed control of the management of all elements the PES.

5.4.6   Documentation

The requirements for PES information is based upon BS IEC 61508-1 Annex A Tables A1 to A3, and the current status of existing documents providing this information is detailed in a matrix within the CMP. This enables the common terminology of information requirements, for example, SRS in BS IEC 61508, to be used throughout the modification process.

However, for legacy systems it is likely that many of these pieces of information are not available, or that shortfalls in the documentation have been found to exist. A pragmatic approach has been adopted which specifies the extent to which missing design documents (for example, User Requirements Specifications) must be reproduced when carrying out a modification based on the Procedural Rigour Level assigned to the PES, as shown in Table 7.

	Non safety	Minor	Significant	Major
Produce full document				✓
Produce part of document (for modification)		✓	✓	Justify
Do not produce a document (use detail in SMR)	Justify	Justify	✗	✗
✓	denotes the level of documentation that should be produced if practicable, for the relevant Procedural Rigour Level.			
Justify	denotes the level of documentation that should be produced as a minimum with justification of why this level of documentation has been chosen, for the relevant Procedural Rigour Level.			
✗	denotes that this level of documentation is not acceptable for the relevant Procedural Rigour Level.			

Table 7 – Requirement to Produce Missing Documentation Versus Procedural Rigour

For example, from Table 7, for a modification to a PES with a 'Major' Procedural Rigour Level:

- A full document should be produced if practicable;
- With acceptable justification, a part document relating to the modification may be produced; and
- It is unacceptable for no documentation to be produced.

As a minimum:

- A boundary diagram has been developed to describe the extent and connectivity of the PES;
- A Safety Requirement Specification containing details of the Safety Function has been produced for all Safety and E,C&C PES; and
- All relevant details supporting the management of the PES is contained in a PES register (SPESR).

6 The Additional Procedural Suite and Supporting Implementation Activities

The project scope required a large amount of supporting work to be carried out in addition to the categorisation and modification process. The main elements of these are summarised in the following sections.

6.1 Operations and Maintenance

Whilst the primary focus of the project was to introduce improved arrangements for the management of modifications the use of Procedural Rigour has had additional benefits in other areas of work. A suite of common arrangements for the operation and maintenance of PES have also been produced, see Figure 13, which also increases in robustness as the Procedural Rigour Level increases, for example, the level of access control applied to systems and amount of failure and demand rate data captured.

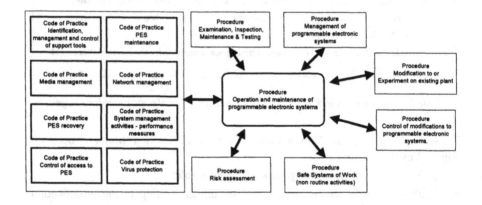

Figure 13 - Operations and Maintenance Procedural Suite

The use of Procedural Rigour Level and system type has allowed for prioritisation of the work, and gives focus to the most important systems.

6.2   Roles, Accountabilities and Competence

The introduction of a common structured process has allowed clear and consistent accountabilities to be assigned to roles within the process. The key roles are the PES Manager and the Plant Manufacturing Manager.

The PES Manager is responsible for the co-ordination of each modification and the technical checking and authorisation of the change.

The Plant Manufacturing Manager is responsible for the safety and operation of the plant and sponsors the modification as the customer and ensures appropriate safety validation is carried out so that plant safety is not compromised.

Twelve other duties were identified within the Operations, Maintenance and Modification processes which could potentially immediately affect safety incorrectly carried out, for example:

- Undertake PES Manager role to ensure allocated PESs are maintained and modified in accordance with the Codes of Practice;
- Performing permanent on-line configuration modifications;
- Performing the role of Functional Safety Assessment Chair;
- Performing the role of Design Review Chair;
- Uploading and downloading software; and
- On-line monitoring/fault diagnosis where access permission allows modification to be performed.

For all of these roles a nomination and appointment process was introduced to assess four key areas of knowledge:

- Technical;
- Plant specific issues;
- Procedures; and the
- Safety aspects of all of these.

Detailed assessment criteria were developed to support the confirmation of the knowledge required. Examples of the questions to support knowledge of the safety functions are:

- Which documents describe the safety functionality?
- What is the safety functionality of the system?
- What is the Procedural Rigour to be applied?
- What are the potential safety impacts of maintenance and modification activities on the system in question?
- What steps must be taken to prevent this?

In addition to specific questions the assessment process also takes into account the practical experience of the candidate.

Key learning – Appointment and recording of competence requires a large time commitment, but, an increased level of knowledge and awareness has resulted from this.

6.3 Training

To support the rollout of the new arrangements, the culture change required and the level of technical nomination a comprehensive suite of 23 training modules was produced. These covered awareness training for all those involved (manufacturing, E&I and control systems personnel) and specific technical training in the use of the process and the methods and techniques to be applied. A total of 7000 people/courses were delivered over the 12 month implementation phase.

Key learning – Training must be developed and delivered for all personnel impacted including these outside the PES community, for example, Plant Managers, Permit to Work Co-ordinators, Safety Case Owners etc, as they are crucial to the success of the project.

6.4 Implementation

The rollout and implementation of the new arrangements was planned and scheduled to take each department from their current arrangements, level of knowledge and supporting systems and to put them in a position to work to the new arrangements. It involved the co-ordination of all aspects of the project and was carried out by a dedicated team to minimise the impact upon the existing teams. A team of over 100 personnel were deployed to achieve implementation.

Key learning – A dedicated project team of 100+ personnel was required to adequately rollout the new arrangements to 1200 PES.

The overall work co-ordinated by the implementation team can be seen in Figure 14.

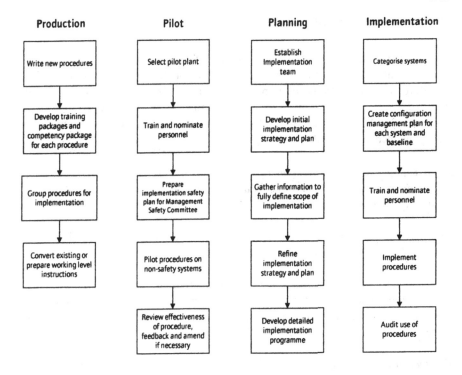

Figure 14 - Implementation Overview process

6.5   Confirmation of the Adequacy of Arrangements

A thorough and comprehensive level of review and challenge to the arrangements was encouraged throughout the production, pilot and approval of the process. This occurred using relevant technical personnel from within BNFL, the use of partners personnel and external consultants. Virkonnen Critical Systems were used to assist in the development and initial assessment of the arrangements, and Praxis Critical Systems were contracted to carry out an independent assessment of the arrangements. Both assessments were conducted against BS IEC 61508 and other industry best practice. All reviews have confirmed the adequacy of the arrangements.

Praxis Critical Systems specifically concluded that *"BNFL can have a great deal of confidence that the overall software modification process is adequate to ensure that software modifications do not adversely affect the safety of a system"*.

6.6   Continuous Improvement

It is recognised that the introduction of new arrangements on this scale will inevitably not be perfect. In order to address this a process for the querying of the procedures and suggestions for improvements is in place. These are routinely reviewed and responses are given to the person raising the query, added to a FAQ list on the BNFL Intranet for access by all interested parties and improvements to the arrangements identified.

Key learning – A team of personnel to support the ongoing use of the procedures, to provide guidance and aid consistent application has been put in place as part of the overall PES organisation.

An improved set of forms is due to be released at the end of 2002 and the key principles reviewed and improved by end of 2003.

Proactive performance measures have been incorporated into the arrangements. These allow routine monitoring and assessment of the effectiveness of the procedures. A programme of co-ordinated internal and external audits are also in place to demonstrate compliance with the process.

Key learning - The project as a whole has cost BNFL greater than £5 million and took over two years to complete. However BNFL is now being rewarded with many indirect and direct benefits.

7   Conclusions

In order to apply BS IEC 61508 successfully to legacy PES a number of key stages have had to be addressed:

1   Identification of all Safety and Safety Related PES, the safety function they fulfil and the dependency placed upon them.

2   Assessment of the PES to demonstrate that it fulfils these requirements.

3   Development and implementation of a quality and technical process based upon the modification and realisation phases of the overall safety lifecycle to ensure that the PES continues to perform at the required level.

4   Introduction and co-ordination of supporting Management of Safety processes, for example, competency and nomination of personnel.

5   Production of practical solutions and guidelines to deal with the discrepancy between legacy PES and those developed in accordance with BS IEC 61508, for example, in the supporting information and specific documentation available.

BS IEC 61508 is a complex standard, which cannot be understood and applied beneficially without significant amounts of review of the document, detailed preparation, and production of specific guidance. Based upon our extensive experience of implementing BS IEC 61508 on 1200 legacy systems we believe the standard is weak in four key areas:

1   The modification model and its relationship back into the overall safety lifecycle.

2   The process for carrying out effective Impact Analysis.

3   The process for carrying out effective Functional Safety Assessment.

4   Knowledge and applicability of the numerous methods and techniques proposed.

In all of these cases the overall principles are declared in the standard but detailed guidance on how to apply them has had to be produced. Considerable resource and consultation were required to produce this guidance at a level that allowed effective implementation to be achieved.

BNFL has maintained the safety of its plants and processes by the use of formal safety cases, risk reduction techniques and the application of a range of processes and procedures. The adoption of BS IEC 61508 and its structured framework has allowed all of these measures to be drawn together into a cohesive overall lifecycle. This has led to the overall enhancement of the management of safety upon PES.

8   Acknowledgements

The implementation of a project of this scope and complexity could not have been accomplished without the skills, knowledge and dedication of a large team of people. Those people all deserve significant credit for the success achieved by this project. They include personnel from within BNFL, particularly the project team members, leaders and managers, and external personnel, such as our Control Systems Partners, Virkonnen Critical Systems, Praxis Critical Systems and our regulator the HSE/NII.

Particular thanks must go to the primary reviewers of this paper; M Litherland, P J Gregory, R Pearce, S Jennings, I Rae, G A Rumney, G R C Maley, K Wheatley and S Wardle.

References

[1] BS IEC 61508:1998 – Functional safety of electrical/electronic/programmable electronic safety-related systems Parts 1-7.

[2] BS ISO/IEC 12207:1995 – Information technology – Software lifecycle processes.

[3] BS EN ISO 9000-3:1997 – Quality management and quality assurance standards - Part 3. Guidelines for the application of ISO 9001:1994 to the development, supply, installation and maintenance of computer software.

[4] HSG65 – Successful health and safety management (HSE: reprinted 2000).

The Software Safety Standards and Code Verification

Ian Gilchrist
IPL Information Processing Ltd
Bath, UK

Abstract

An analysis is offered of the current principal safety-related software standards with regard to their requirements for software verification and testing. The principal points of agreement are presented, and attention is drawn to differences. These differences occur most significantly in the areas of recommended techniques to be used during (unit) testing, and the basis for decision as to what constitutes 'enough' testing. Some mention is also made of the differing roles of static analysis as a verification technique. The reader is invited to speculate on the effects of these differences, and the question is raised as to whether too much testing is being demanded or too little.

Introduction

Software verification activities, especially testing, occupy a large part of the development cycle for safety-related systems development. All the applicable standards agree that these activities are an essential part of the overall process of creating certifiably safe systems. This paper is an attempt to analyse what is common to the principal standards, and also to focus on some of the details where less agreement is evident.

The available evidence suggests that, when followed properly, the current standards do work. That is to say, there have been few public examples of certified systems failing due to a breakdown in the code verification process. But given the lack of agreement on what constitutes a really 'safe' approach to software verification it might be wondered whether this is just a matter of luck. At the end of a survey like this, one is left wondering what the engineering basis for some of the currently used or recommended techniques is.

There is the further question of whether or not projects are being required to do too much or too little verification. The consequences of too little testing are obvious and are of great concern to all professionals involved in software safety. However, it is also reasonable to ask whether or not current practices veer too much the other way. Too much testing can involve a significant cost burden and this might be avoided if agreement could be reached on how *best* to achieve an acceptable level of confidence in safety-related software.

The Safety-Related System (SRS) Standards

Table 1 defines the standards that currently constitute the state of play in our industry, their origination and use.

STANDARD	Originating Body	Use
[Def Stan 00-55]	UK Ministry of Defence (MoD)	UK MoD defence projects
[RTCA DO-178B]	Radio Technical Commission for Aeronautics (RTCA)	International civil avionics systems
[IEC 61508]	International Electrotechnical Commission (IEC)	International generic standard for all Programmable Electronic Systems
[PrEN 50128]	European Committee for Electrotechnical Standardization (Cenelec)	Rail industry (derived from [IEC 61508])
[IEC 601]	IEC	Medical equipment (derived from [IEC 61508])
[FDA Guidelines]	US Food and Drug Administration (FDA)	Software Validation and Verification (V&V) for medical equipment
[IEC 880]	IEC	Nuclear power
[MISRA Guidelines]	Motor Industry Software Reliability Association (MISRA)	Motor industry
[MISRA Safe C]	MISRA	Use of the C language in vehicle-based software.

Table 1. Current software safety standards, their source and use.

What Broad Principles Do The Standards Agree On?

There are a number of issues on which the majority of the standards agree.

Systems are assigned a 'safety level'. Part of the preparation prior to starting a safety-related system development is the assignment of a Safety Integrity Level (SIL). For most of the IEC standards this tends to be numbered 1-4, with SIL 4 representing the safety-critical end of the spectrum. The main exception is [RTCA DO-178B] which has Levels A-E, with Level A being the safety-critical end. These SILs are important to developers because they tend to determine the level of rigour to be employed in the various development activities including verification and testing. Generally speaking, the higher the SIL the more onerous (and hence expensive) the verification demands are.

Verification requires a combination of techniques. Developers should not rely exclusively on one technique to gain confidence that their software works properly. In very broad terms the main techniques recognised are *code reviews, code analysis,* and *testing.* For example, [IEC 61508] says that, "It is the combination of code review and module testing that provides assurance..." [Def Stan 00-55] is well-known for its favouring of formal analysis methods, but does allow, "Where full formal verification... has not been performed the amount of testing will need to be increased."

Testing on its own is not proof. The purpose of testing is, "to discover errors, not prove correctness" [MISRA Guidelines]. It is accepted that no amount of testing can conclusively show that there are no more bugs. Further, testing on its own cannot demonstrate that a given piece of code is safe. So testers need to try to create conditions which would show up faults if they are present. This limitation in the power of testing is the main reason for the need for a combination of techniques.

Verification activities are best done by an independent team. The purpose of all verification activities is to show that the 'implementers' have correctly created what the designers intended. For implementers to be involved in verification risks weakening the rigour that a safety-related system needs. For example, [MISRA Guidelines] states, "it is recommended that V&V are carried out by a person or team exhibiting a degree of independence from the design and implementation function."

Traceability. It is important that all requirements can be traced down through design and then into implementation and verification. This is the preferred route for ensuring that all requirements are actually incorporated and correctly so. The different levels of testing give a number of opportunities to verify that any specific requirement is included. Most of the standards ask for some kind of traceability matrix as part of the overall safety case. [RTCA DO-178B] says, "The verification process provides traceability between the implementation of the software requirements and their verification."

Test Environment. The target platform for a system is not necessarily the same as the development platform, so the question of where testing should be done arises. Preference is stated for running tests on the target, but most standards make allowances for doing testing on a simulated target or even the development platform. [Def Stan 00-55] says, "The most valuable form of testing is conducted on the real target hardware using the real code..." but that, "It may be desirable to pre-run the tests on a host (to debug the code for test coverage)."

Repeatability. There is agreement among the standards for a testing regime which allows tests to be re-run on demand. In part, this comes from the need for an independent test team but it is also sensible when considering the development process: in a typical lifecycle, code will need to be tested and retested a number of times before it can be shown to 'work'. [Def Stan 00-55] states that, "The test environment... shall be recorded in a manner... which allows the tests to be re-run."

Testability. Some of the standards are prescient enough to recognise that testing needs to be considered at the design stage. [IEC 61508] says, "The software should be produced to achieve testability," while [MISRA Guidelines] suggests developers should, "design for verification and validation bearing in mind the capabilities and limitations of the available testing tools, rigs..."

Levels of Testing. The well-known V-Model of the software lifecycle suggests staged verification activities at (in general terms) the Unit, Integration and System/Acceptance levels. This concept is widely endorsed by the standards, but in fact all they can agree on is that it is up to a project's engineers to determine what lifecycle model they will use and consequently what levels of verification they will do.

The demands of the standards at the various levels are now developed in the following sections:

Unit Testing

What is Unit Testing? Recognising that software systems are composed of modules, unit testing is the verification that these modules do what they should do. [IEC 61508] puts it very succinctly: "Each software module shall be tested as specified during software design. These tests shall show that each software module performs its intended function and does not perform any unintended functions." Unit testing is a requirement of all the standards for the higher SILs, but may be omitted for lower SIL projects, at the discretion of the project engineers and if suitably justified.

It should be noted that merely performing unit testing is not sufficient. [IEC 61508] helpfully reminds us that, "The results of software module testing shall be documented," so as to form part of the safety case evidence, and also that, "The procedures for corrective action on failure of a test shall be specified."

The basic elements of unit testing are well established. A software component has 'inputs' (such as procedure parameters and global data). Prior to execution of the unit, the tester will select appropriate input values and predict the corresponding outputs the unit will generate based upon the specification. A well-run, thorough testing activity will ensure that the actual and expected outputs are compared in a systematic, automated fashion. This will greatly aid the goal of repeatability.

The following issues also need to be considered when planning unit tests:

- Units are rarely testable in isolation. It is almost always necessary to provide some simulations for at least some of the external interfaces of the unit under test. A variety of techniques exist for enabling this, the best known of which is 'stubbing'. Careful consideration needs to be made when determining what software is stubbed and what 'real' software is included in each unit test definition.
- It is often useful to monitor the internal states of the unit under test. This kind of testing is usually referred to as 'white-box' testing. Various (language-dependent) techniques exist for doing this.

The decisions made regarding these issues should be recorded as part of the project's overall Software Verification Plan [RTCA DO-178B].

(Unit) Test Case Definition. Whilst there is good general understanding of what unit testing is there is a certain amount of disagreement on how units should be tested in order to gain an adequate degree of confidence that they work reliably. A wide range of possible techniques are mentioned in the various standards, but there is only moderate commonality, and the theoretical basis for some of the techniques is questionable. Table 2 gives some idea of the extent of the commonalities and variations.

STANDARD	Functional Testing	Structural Coverage	Equivalence Partitioning	Boundary Values	State Transition	Robustness	Error Seeding
Def Stan 00-55	Yes	Yes	Yes	Yes	Yes	Yes	
RTCA DO-178B	Yes	Yes	Yes	Yes			
IEC 61508	Yes	Yes	Yes	Yes		Yes	Yes
PrEN 50128	Yes	Yes	Yes	Yes			Yes
IEC 880	Yes	Yes	Yes	Yes		Yes	
MISRA Guidelines	Yes	Yes	Yes	Yes		Yes	Yes

Table 2.Comparison of recommended unit testing techniques.

Some of the standards go to the trouble of distinguishing between techniques as (Highly) Recommended or not according to the system's SIL. Table 3 lists these for [IEC 61508]:

SIL	Boundary Values	Error Guessing	Error Seeding	Performance Modelling	Equivalence Partitioning	Structural Coverage
4	HR	R	R	HR	HR	HR
3	HR	R	R	R	R	HR
2	HR	R	R	R	R	R
1	R	R	-	R	R	R

Table 3. IEC 61508 recommended test techniques.
'HR' = Highly Recommended; 'R' = Recommended

How Much Testing Should be Done? The [FDA Guidelines] state this issue clearly: "A developer cannot test forever," so, "it is a matter of developing an acceptable level of confidence." The standards differ markedly on both the forms and amount of testing suitable for safety-related software. However, the definitions provided are frequently vague and rely heavily on users' engineering judgement. The most common recommendation appears to be a combination of ensuring that all 'requirements have been met' and the use of code coverage analysis. The former stipulation can be met laboriously (and subjectively) by the construction of a traceability matrix. Code coverage has two advantages: it can (mostly) be automated, and it is objective. However beyond those points there is again little agreement. For example, [Def Stan 00-55] says that, "All source code statements and branches should be executed"; [MISRA Guidelines] says that, "Test coverage should be considered carefully"; [RTCA DO-178B] states that the amount of coverage needed will depend on SIL – see Table 4.

SIL	High-Level Requirements	Low-Level Requirements	Statement Coverage	Decision Coverage	MC/DC* Coverage
Level A	SI	SI	SI	SI	SI
Level B	S	S	SI	SI	-
Level C	S	S	S	-	-
Level D	S	-	-	-	-
Level E	-	-	-	-	-

Table 4. RTCA DO-178B Test Coverage Requirements.
'SI' = Satisfy with Independence; 'S' = Satisfy.
*MC/DC is a form of Boolean Operator or Condition/Decision coverage.

Other forms of coverage may be required as well. [Def Stan 00-55] for example has a long list, which includes the statement and branch coverage requirement already mentioned, with the additional requirements to set all source code variables to min, max and intermediate values, all Boolean and enumerated variables to all possible values, all loops to be executed 0,1, max and intermediate times etc.

It is worth pointing out that even the highest level of coverage in [RTCA DO-178B], which in theory ought to give a high degree of confidence that most faults have been found, by no means finds all. Recent research by Bristol University showed, in a single sample of code, a significant level of residual faults even after 100% Modified Condition/Decision Coverage (MC/DC) coverage had been achieved [Kuball et al.]. A further issue with MC/DC coverage is that the possibility of achieving full coverage can be language dependent: as originally formulated MC/DC could be achieved using the Ada language but not with C. The reason for this is the latter's use of condition short-circuiting which makes evaluation of certain combinations of condition which MC/DC requires impossible to achieve.

Static (Code) Analysis. This is another area of great disagreement between the standards. This is not surprising considering the great range of techniques which the title encompasses. Here is a small sample of some current uses to which static analysis is put:

- To enforce bans on certain code constructs considered 'unsafe'. There is no agreement on what should be in this list, probably because it is highly language specific. Sometimes code constructs are banned because they are considered intrinsically unsafe, and sometimes because their use prohibits more complex forms of analysis.
- To enforce the following of coding standards which are considered 'safe' and the avoidance of those considered 'unsafe'. The [MISRA Safe C] subset is the best example of this. Possibly the best that can be said is that this represents a reasonable start to formalising the use of the language in a safety-related context.
- To follow certain ideas on what constitutes safe and (formally) correct coding. The types of techniques here include Control Flow, Data Flow and Information Flow analyses. Spark Ada is probably the best example of a proprietary analyser supportive of these techniques.

Integration Testing

Variations in Integration Testing. There is some considerable variation in approach and terminology used in the various standards when integration testing is discussed. Table 5 lists some.

Software engineers will recognise the following possible levels at which integration testing *might* be done: cluster testing (groups of units/modules, which corresponds to the term 'software integration testing' above), task or thread testing (has no simple correspondence with terms above), hardware-software integration testing (has easily identifiable correspondence in the table), and sub-system testing (typically multiple tasks). Depending on the size and SIL of the project concerned it may be justifiable to leave out some or all of these levels. [Def Stan 00-55] suggests, "It may be possible to omit integration testing provided that the software interfaces can be shown to be exercised by the system tests." [RTCA DO-178B] would require that whatever decision was made it would need to be logged in the project 'Software Verification Plan', along with the justification for inclusion or omission.

Standard	Identified Integration Test types
RTCA DO-178B	Software Integration tests Hardware-Software Integration tests
Def Stan 00-55	Integration tests
IEC 61508	Software Integration tests Component and Sub-system Integration tests
IEC 880	Hardware-Software Integration tests
MISRA Guidelines	Full Software Integration tests Hardware-Software Integration tests ECUs integrated into vehicle

Table 5. Terms used to describe different approaches to integration testing.

It is reasonable to ask at this point what the purposes of integration testing are. Not surprisingly this depends on the standard to be complied with and SIL. Table 5 lists some of the stated purposes for both [RTCA DO-178B] and [Def Stan 00-55]. It is sometimes not clear whether the standards expect all of these things to be done, or just some, and if so how they should be done. It is worth contrasting the differing levels of detail in the two standards, in Table 6.

RTCA DO-178B 'Software Integration testing'	Def Stan 00-55 'Integration testing'
"Incorrect initialisation of variables." "Violations of software partitioning." "Interface or parameter passing errors." "Data corruption especially global data." "Inadequate end-to-end resolution." "Incorrect sequencing of events and operations."	"Shall as a minimum demonstrate the correctness of all interfaces."

Table 6. Stated purposes of RTCA DO-178B and Def Stan 00-55 'Software/Integration testing'

System Testing

System testing is really outside the strict scope of any discussion of code verification, since it is mainly about ensuring that all (system) requirements as specified have been met. This may include testing on aspects such as timing, availability, capacity etc. Some standards will specify details of what is expected in addition to this basic requirement. For example, [Def Stan 00-55] says, "All numerical inputs and all outputs (should be) set to their minimum, maximum and an intermediate value;" and more in similar vein.

Opinions seem to differ whether safety can be adequately demonstrated by testing. Some of the standards talk about Validation testing. For example, [Def Stan 00-55] says, "Validation testing shall be performed to demonstrate that the SRS operates in a safe and reliable manner..." and [IEC 61508] talks about, "Safety Validation...(checks that)... the integrated system complies with the specified requirements for software safety at the intended SIL."

Conclusions

Whilst there is a reasonable amount of agreement amongst the safety-related standards about the principles of good verification and testing practice, there is significant variation especially regarding details. The overall objective evidence suggests that the standards 'work' as there have been few failures involving software where a certified process was followed. However the disparity of techniques, both mandated and recommended, suggests that there is still lacking a scientific basis for their use. This can lead to three possible notions:
1. That the success so far achieved has been a matter of luck and that it is just a matter of time before something does go wrong.
2. That over-engineering is being required, leading to an unnecessary cost burden. A more scientific approach to determining 'value for money' on the verification side *might* allow costs to be cut, but it would take a brave person to start down that line.
3. That the proper course of action is to tighten still further the verification requirements, as there is evidence that even at the highest levels bugs are still getting through.

Web Site References

RTCA. Radio Technical Commission for Aeronautics. (see www.rtca.org)
Def Stan. (UK) Defence Standard (see www.dstan.mod.uk)
IEC. International Electrotechnical Commission. (see www.iec.ch)
Cenelec. European Committee for Electrotechnical Standardization. (see www.cenelec.org)
FDA. (US) Food and Drug Administration. (see www.fda.gov)
ISO. International Organisation for Standardization. (see www.iso.ch)
MISRA. Motor Industry Software Reliability Association (see www.misra.org.uk)
Spark Ada. A formally provable subset of the Ada language (see www.praxis-cs.co.uk).

Document References

Def Stan 00-55. Requirements for Safety-related Software in Defence Equipment, Issue 2. Ministry of Defence, UK, 1997

FDA Guidelines. General Principles of Software Validation: Final Guidance for Industry and FDA Staff. Food and Drug Administration, USA, 2002

IEC 601. Part 1 General Requirements for Safety, 4. Collateral Standard: Programmable Electrical Medical Systems, International Electrical Commission, 1996-05

IEC 61508. Functional Safety of Electrical/Electronic/Programmable Electronic Safety Related Systems. International Electrotechnical Commission, Geneva, 1998 – 2000

IEC 880. Software for Computers in the Safety Systems of Nuclear Power Stations, International Electrical Commission, 1986

Kuball S, Hughes G and Gilchrist I. Scenario-Based Unit Testing For Reliability, Proceedings of the Annual Reliability and Maintainability Symposium RAMS 2002, Seattle, USA, IEEE 2002

MISRA Guidelines. Development Guidelines for Vehicle-Based Software. The Motor Industry Research Association, UK, 1994

MISRA Safe C. Guidelines for the Use of the C Language in Vehicle-Based Software, Motor Industry Research Association, UK 1998

PrEN50128. Railway Applications - Software for Railway Control and Protection Systems, Draft pr EN 50128, CENELEC/SC 9XA, Dec 1995

RTCA DO-178B. Software Considerations in Airborne Systems and Equipment Certification. RTCA Inc, Washington USA, Dec 1992

Evolution of the UK Defence Safety Standards

J A McDermid,
Department of Computer Science, University of York,
Heslington, York YO10 5DD

Abstract

The MoD has a range of safety standards, some of which have been in use since the early 1990s. There have been suggestions for change to the standards based on industrial experience of using the standards and in response to the MoD's Standards Breakthrough process. This paper summarises the reviews of the standards and outlines current proposals for reworking DS 00-56, the main system safety standard.

1   Introduction

Most safety critical system development is guided or constrained by standards. Standards vary between industrial sectors and between countries. Standards also evolve over time to reflect changes in technology, changes in understanding of the safety process, and modifications in the licensing or regulatory process, where there is one. Several standards have been developed for use in the UK Defence sector; this paper gives an overview of the development and evolution of these standards, focusing on recent proposals for updating the key standards.

The UK has a number of Defence Safety Standards. DS 00-56 (MoD 1996b) sets out a framework for system safety engineering. There are three technology-oriented standards. DS 00-54 (MoD 1999a) deals with hardware. DS 00-55 (MoD 1997) and DS 00-58 (MoD 1996a) deal with software. DS 00-58 and DS 00-54 are relatively recent publications. There is little experience of their use, and they are still at their original issue; at present there is little pressure to update them.

DS 00-56 and DS 00-55 were originally issued almost a decade ago, and they were later upissued to reflect experience in their use, and maturing of understanding of safety engineering. The current versions have been in force since 1996 and 1997 respectively. There has been considerable experience in using these two standards, or parts thereof, on defence projects. Whilst there have been positive experiences in using the standards there have also been some difficulties, and a measure of controversy, since the standards were first introduced. For example, there has been some concern at what was perceived to be over-emphasis on formal methods in DS 00-55. Also, aspects of DS 00-56 have been challenged as it was initially focused solely on at electrical, electronic and programmable electronic systems (E/E/PES) but the second issue of the standard removed this restriction – albeit with little modification to the content to make it fully suitable for this broader role.

Industry comments and feedback on the standards prompted a number of reviews of their contents, and suggestions for reworking the standards. Recently the MoD set up a Safety Standards Review Committee (SSRC)[1] to oversee the review of the standards and to direct any necessary rewriting. At the same time the MoD has embarked on a Standards Breakthrough process intended to reduce the number of defence standards, and to move towards the use of civilian standards, wherever appropriate. This is analogous to the Perry initiative in the USA, although the targets for reducing the number of defence standards is less aggressive.

Although the SSRC is doing some work on an update to DS 00-55, it is currently focused on DS 00-56 as this is the "umbrella" standard, providing a context for the remaining defence standards. Section 2 outlines the work in the SSRC and other activities which are influencing the development of DS 00-56, including giving brief discussions of the reviews that pre-dated the establishment of the SSRC. Section 3 outlines the requirements for, and outlines some of the key issues being addressed in producing,, DS 00-56 Issue 3. Section 4 presents some conclusions and outlines the next steps in evolving the standards.

2   Evolution of DS 00-56

There are four main strands of activity influencing the development of the defence standards, especially DS 00-56. Responsibility for the standards was originally with the MoD's Equipment Safety (Policy) (ES(Pol)) branch, before it was disbanded. ES(Pol) launched the initial studies into the defence standards, and it was several years before the other initiatives started. There are now three parallel, and inter-related, activities. We discuss the work supported by the SSRC last as this makes it easiest to explain how the other work is influencing the body which is formally responsible for the evolution of the standards.

2.1   ES(Pol)

ES(Pol) funded two studies of the standards, focusing on DS 00-55 and DS 00-56. One was undertaken by Viv Hamilton and Ceri Rees of the GEC Marconi Research Centre at Great Baddow. The study other was undertaken by the author.

The GEC Marconi work concentrated on drawing lessons from industrial experience of applying the defence standards, particularly within their own organisation. It also considered technological changes which affected the standards. The second study focused more on technical issues with the standards, although it drew on industrial experience, e.g. through work as an independent safety auditor (ISA).

---

[1] The author is a member of the SSRC, but the views expressed in this paper are those of author, not the SSRC or MoD.

The detailed technical recommendations of these studies[2] are rendered somewhat obsolete by the passage of time and, to some degree, by the MoD's Standards Breakthrough process, so they are not repeated here. However, there was one key finding about DS 00-56, which remains pertinent.

Both studies found that there was a need for an "over-arching" standard which covered all aspects of system safety, i.e. applied at the platform or whole (weapon) system level. In particular the two studies concluded that there were omissions from DS 00-56 if it was to serve as a standard at the platform/complete system level, rather than the E/E/PES level. Further it placed requirements which were not appropriate at platform/complete system level. However, the reports also noted that any new standard could draw heavily on the then current version of DS 00-56, as the core safety management requirements are contained in the standard, and are quite general.

2.2   DARP

The Office of Science and Technology established the Foresight programme to help guide research in a range of industries. The Defence and Aerospace Foresight Programme decided in 1998 that it was appropriate to establish and recognise national centres in key areas of technology. These centres are known as Defence and Aerospace Research Partnerships (DARPs). The first DARPs were recognised in 1998, and there are now more than ten of these centres in various technologies.

The High Integrity Real-Time Systems (HIRTS) DARP was founded in 1998. It was established as a partnership between BAe (now BAE SYSTEMS), DERA (now QinetiQ), Rolls-Royce and the University of York. The HIRTS DARP has run a series of workshops involving the partners, the MoD, system suppliers such as Smiths Industries, and certification agencies such as the CAA. The defence safety standards were discussed at the first workshop in April 2000, and again at the workshop on April 30th to May 1st 2002. The results of these discussions can be found on the DARP website (HIRTS DARP).

One of the key outputs of the April 2000 workshop was a SWOT analysis of the standards, see overleaf. The SWOT analysis is "broad brush", but gives a feel for the positive aspects of the standards, and the concerns which the standards cause in the community. This workshop also concluded that it was appropriate to produce an "over-arching" standard, and gave consideration to the role of existing standards (including those from outside defence) in the context of this top-level standard. An outline "map" was produced, showing non-defence standards and other MoD documents, e.g. the main Joint Service Publications (JSPs) in the framework.

---

[2] Because the bulk of these reports is now out of date, and they are strictly not in the public domain, references are not given.

Strengths

- Under UK control
- They are standards
  - usually know what they mean
- Encouraged
  - suppliers to improve practices
  - suppliers to demonstrate system safety
  - research and technology transfer
- Sound underlying principles

Opportunities

- Standards Breakthrough
  - defence standards when no commercial equivalent
  - remaining standards better managed
- International harmonisation
- Benefits of consensus
- More "evolvable" standards
- Focus on safety
- Setting targets which we can measure against
- Usable by competent engineers not "super-heroes"
- Common standard
- Remove ambiguity

Weaknesses

- Not internationally recognised Mixed authorship mean that parts are hard to understand
- Only "four of many" and conflicts with other standards
- Defence standards
- No known project that complies fully
- Uninformed insistence on compliance by MoD requirements managers
- Not "user friendly" to: legacy, International projects, COTS, NCOTS, SOUP, Commercial situation

Threats

- Standards breakthrough
- Initiative "overload"
  - too much change
- Weakening standards to get harmonisation
- Lack of visibility to decision makers
- Effects of strong standards
  - phase lag
  - standards legacy
  - damage to competitiveness
- Lose drive for technology
- Hindrance to system development
- Results not used (NIH)

Table 1: SWOT Analysis of Defence Safety Standards

The second workshop in 2002 built on the initial ideas set out in 2000, the work of the SSRC and that supported by QinetiQ (see below). The primary focus was a discussion of the proposed requirements for DS 00-56 Issue 3. These requirements are discussed in section 3.1, so they will not be repeated here. The meeting suggested some detailed changes to the requirements, but were broadly supportive of the proposal – indeed one of the industrial attendees said that the proposed approach had the potential to be "world class" if implemented properly. Several attendees said that it would be much easier to judge the validity of the requirements given a proposed (draft) text for the standard – i.e. they wanted a design not requirements! One of the most common observations was a need to "get on with it", as there was a feeling that relatively little progress had been made in two years.

2.3   QinetiQ

QinetiQ's Systems Assurance Group (SAG) runs a number of research projects in the area of safety critical systems. One MoD-funded project is studying technology associated with software safety assurance, and has an explicit objective to influence the development of the defence safety standards. Three activities are relevant to the evolution of DS 00-56.

First, a study was undertaken to investigate ways of allocating safety requirements to systems and software, particularly trying to demonstrate that driving safety analyses down into the software design was more effective than a SIL-based approach. The study involved the Software Verification Research Centre (SVRC) in Brisbane, the University of York and MBDA. The research was based on a based on a "model" of a missile control system, provided by MBDA. The work was started at a workshop was held in York, and was completed by Peter Lindsay of the SVRC. The main ideas developed in this study have since been published (Lindsay 2001) and have influenced some aspects of the draft standard. In particular the study gave confidence that it is appropriate to ask for derived safety requirements to be produced by the safety process, and allocated to software.

Second, a workshop (McDermid 2001) was held at Puckrup Hall in Tewkesbury in between the two DARP workshops. This involved most of the SSRC, other interested parties in the MoD, staff from QinetiQ and the author. Also, some overseas experts were invited to give an international view on the development of DS 00-56. The workshop produced a number of important conclusions:

- The Standards Breakthrough process requires civilian standards to be used, if they are appropriate. It was agreed that there is no suitable civilian standard, as none address the safety issues faced by weapons systems suppliers, and none currently interface to the JSPs e.g. JSP 430 (MoD 2001). Also, it is unlikely that any civilian standard would fill this role in the future. A primary reason is that there is no reason for a civilian standard to interface to the JSPs, as they are only applicable to military systems.

- The Standards Breakthrough process also asks whether or not other military standards can be used. The US MilStd 882D (DoD 1999) is a much reduced version of the earlier 882C (DoD 1993). It was generally accepted that MilStd 882D is too weak to serve as a basis of UK procurement, as it doesn't cover key issues, e.g. production of a safety case, which are fundamental to UK practises (and are called up in JSPs and the MoD's Acquisition Management System (AMS)).

- The new standard should be DS 00-56 Issue 3, and should be written as a "contracting version of the JSPs, and relevant aspects of the AMS".

In addition, the workshop identified key stakeholders, both within and outside the MoD, who needed to be consulted to get acceptance of a new standard. These include the CAA, the HSE, and professional bodies such as the BCS and IEE.

Finally, QinetiQ are funding the development of the draft of DS 00-56 in support of the SSRC. The status of this work is discussed in section 3.2.

2.4 SSRC

The SSRC is an MoD group involving representatives of the different MoD stakeholders, including the various domain-specific Safety Offices, with the author as an outside advisor. The group is run by the Ship Safety Management Office (SSMO) who are the sponsor of the Defence Standards. The group's central remit is to *review* the standards in order to decide if further development work is needed. In other words it was not a foregone conclusion, when the SSRC was established, that DS 00-56 would be rewritten, although this need has now been accepted.

The SSRC has guided the work on DS 00-56 Issue 3 for over two years. Some of the most important activities have been identifying and agreeing the strategy for managing the evolution of the defence safety standards, including DS 00-56 Issue 3, particularly the right way to respond to the Standards Breakthrough initiative.

The SSRC has also sponsored a number of key studies, including the development of advice on safety of Software of Uncertain Pedigree (SOUP). This guidance is now available on the MoD intranet as part of the advice on implementing the AMS. In the context of this paper, probably the most significant study has been an analysis of users' experience with the defence standards. This work was funded through the SSRC and undertaken by Frazer-Nash Consultancy Ltd.

Frazer-Nash produced an extensive questionnaire about DS 00-55, 00-56 and 00-58. The questionnaire addressed factual issues regarding the standards, presentational issues and, most importantly, practical experience of working to the standards. The questions also covered broad issues such as the principles underlying the standards and their relationship to other (international) standards.

Frazer-Nash collected information from about 20 different companies as well as other organisations and individuals, in many cases receiving multiple responses from the same company. This was followed by an extensive task of collating and analysing the responses, both highlighting particularly important observations, and assessing the "balance" of opinion. With such a survey it is not practical to do a true statistical analysis but the responses were analysed to identify which principles or clauses in the standard were broadly supported by the community, and which caused difficulty. This was then used to make judgements about whether or not each standard as a whole needed revision, or reconsideration.

DS 00-58 was found to be fairly uncontentious, and generally liked by those who had worked to it. The general consensus (of those interviewed) was that it was no longer necessary for it to be a defence standard, and it was more appropriately viewed as a guidance document. A final decision on how to reposition DS 00-58 has yet to be taken, although the SSRC accepted the study's main findings about this standard. However the Standards Breakthrough process would suggest that the most appropriate route is to seek its adoption by a civil standards body.

There was considerable support for the principles underlying DS 00-55 – perhaps more than would have been expected given its unusually strong emphasis on formal methods. Some technical observations were made on the standard, but there were also a lot of observations about project and economic issues associated with use of the standard. Probably the greatest concern was about the perceived costs of complying with the standard, and the impact that this has on UK competitiveness. It was also noted that this perception leads to many projects avoiding safety critical software whenever they can (this applies to the MoD, as well as contractors). The SSRC accepted the broad findings of the study, and agreed that it was more appropriate to consider work on DS 00-55 in detail once the position of DS 00-56 had been resolved.

The views on DS 00-56 were mixed. For example, some companies had used SILs with considerable success, whereas others had found them problematic. Some of the most insightful comments came from companies, such as Adelard, who have worked as ISAs and have seen some of the problems of interpreting the standards in a range of organisations. For example, various ISAs had seen difficulties caused by a lack of understanding on how to adapt HRI tables for use by a particular project from the example in the standard. It was noted by many that insufficient guidance was given on such issues – although a significant minority pointed out that a standard cannot be a tutorial or a "distance learning system". Many of the comments reflected the SWOT analysis above, commenting on the difficulties imposed by the standard in an international project. Some respondents also made suggestions on how to resolve the problem, including the adoption of IEC61508 (IEC 1999) – but see below.

Although the survey was by no means universally critical of DS 00-56 it was agreed by the SSRC that there were enough significant issues raised that it was appropriate to rewrite the standard, to respond to these comments and to bring it into line with the Standards Breakthrough initiative. The requirements for Issue 3 of the standard is being produced are outlined in section 3.1 below.

The full Frazer-Nash report is not publicly available, but a summary is (Frazer-Nash 2001).

At the time of writing Frazer-Nash have just completed a series of interviews to seek out the views of industry and other stakeholders on the requirements for Issue 3 of DS 00-56. This will be used to help shape the draft of Issue 3, and to "prime" people on what to expect in Issue 3 before consultation starts on the draft standard.

3   Proposal for Issue 3 of DS 00-56

At the time of writing there have been two major activities relating to DS 00-56 Issue 3. First, a succinct requirements statement was produced to try to get agreement to the approach to rewriting the standard. Some initial feedback was obtained, from the SSRC prior to discussion at the DARP. As the DARP feedback was broadly positive, it was decided to work on a draft of Issue 3, in parallel with the Frazer-Nash industry consultation. Section 3.1 discusses the requirements, and section 3.2 outlines the current activities on rewriting the standard.

3.1   Requirements

A short requirements document (roughly two sides of A4) was written early in 2002 to give a basis for consulting with the MoD and industry on the direction to be taken in rewriting DS 00-56 (as outlined above). These requirements are not reproduced in full here, but the key objectives are set out below, together with a summary of the technical requirements:

- Main objective – a standard which can be called up in contracts to ensure that suppliers provide systems or services which meet MoD safety policy, and enable the MoD to comply with the requirements of the AMS and JSPs.

- Key subsidiary objectives were:

  - The standard should be applicable in all Acquisition scenarios, e.g. equipment provision, service provision, or when using a Private Finance Initiative (PFI);

  - The standard should set requirements with the minimum of constraints on how to meet them;

- The standard should meet the MoD's policy that standards should be "as civil as possible and only as military as necessary";

- The standard should cover environmental issues as well as safety of personnel and material;

- The standard should be usable in international Acquisition.

- The main intent of the standard is to reduce safety and project risks throughout the Acquisition life-cycle, especially use of a system or equipment in service.

- There were fourteen "technical" requirements, covering the classical safety management activities, e.g. identifying hazards, operating a hazard management system, producing safety cases, and so on.

Some observations are in order. First, it is interesting to ask to what extent a standard could meet the "as civil as possible and only as military as necessary" objective. The aim was that DS 00-56 Issue 3 would impose minimum requirements, some of which could be met by using other standards, e.g. ARP 4754 (SAE 1996a) and 4761 (SAE 1996b) in aerospace, where they could be shown to meet the intent of the standard. In other words that the main requirements of DS 00-56 Issue 3 would be "as military as necessary", but could be met, at least in part, by civilian standards, and thus be as "civil as possible".

Second, there was an explicit requirement to keep the standard brief – ideally under 20 pages. Part of the motivation was to make the standard short enough it could be read by project managers and IPT Leaders. Another motivation was to avoid writing a tutorial – and one of the perceived difficulties with DS 00-56 Issue 2 is that it is in part a standard, and in part a textbook, if not a tutorial. Nonetheless it is clear that support material would be needed to assist users in understanding the requirements of the standard and to implement it effectively (see section 3.2 below). There is now much more supporting material, e.g. the MoD's JSPs, which should make it easier to provide this support to the standard, but no firm decision has yet been reached on how to produce this additional material.

Third, in order to be internationally usable the standard should be compatible with MilStd 882. Given the current emasculated form of 882, i.e. issue D (DoD 1999), this should not be a difficult requirement to meet! There has been pressure from some quarters to adopt IEC61508 as (the basis for) DS 00-56. However IEC61508 is aimed at E/E/PES (as was the original DS 00-56), not platforms or complete (weapons) systems as required for the new issue of the standard. Thus, whatever its merits or demerits, IEC61508 is of an inappropriate scope to be the top-level standard for the MoD. However, this does not stop it from being one of the (civil) standards used to show partial compliance with DS 00-56, see below.

Finally, perhaps controversially, there is no mention of SILs in the technical requirements. Instead the focus is on setting risk reduction requirements and showing that they have been met. This doesn't stop organisations from using SILs if they find them useful, but avoids making this a mandatory requirement.

3.2 Status of the Draft Standard, and Key Issues

At the time of writing a first draft of DS 00-56 Issue 3 has been produced for review by the SSRC, prior to producing a revised version for wider review and assessment. The initial draft was produced by the author to "get the ball rolling" but the main responsibility for drafting the standard now lies with the SSRC, with the author continuing to provide input.

As the draft standard is still evolving it would be inappropriate to provide details of the current draft. Instead we give a brief overview of the structure of the standard, and highlight some of the key challenges in producing a standard which meets the above requirements[3].

The current draft contains a small number of "shall" statements – only 3½ pages, and about 12 pages of "should" statements. The aim is to *mandate* the minimum necessary for a contractor to satisfy MoD policy, to allow contractors to use their own existing processes where appropriate, but to give guidance on how to meet this necessary minimum for contractors who do not have suitable existing processes. This is summarised in figure 1 below.

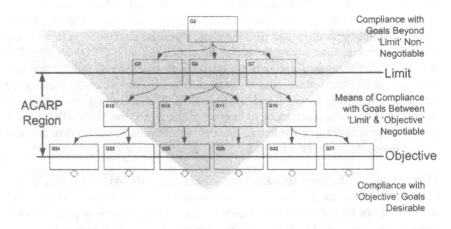

Figure 1: ACARP Principle (Courtesy of Tim Kelly)

---

[3] It is the intention that drafts of the standard will be made publicly available before or at SSS03 and that the presentation will discuss the standard in more detail in order to obtain feedback on the details of the standard.

My colleague Tim Kelly introduced the notion of the "As Compliant as Reasonably Practicable" (ACARP) principle, shown in Goal Structuring Notation (GSN) in Figure 1. In this figure the "mandatory minimum" requirements are the Limit goals, and the "guidance" for contractors without their own existing processes is the Objectives. The Limit corresponds to the shall statements, and the Objectives to the should statements. In writing the standard it is not difficult to mandate the Limit, but one of the challenges is to ensure that contractors meet the Objectives, either by using their own processes or the detailed clauses of the standard. One approach is to ask them to provide an ACARP argument. To try to "flesh out" this idea, aspects of the current draft are being represented in GSN to show how it reflects the ACARP principle, and thus (hopefully) to provide confidence that the standard is strong enough to ensure that contractor use effective safety processes, without imposing undue constraints.

It is intended that DS 00-56 will be is the top of a hierarchy of documentation on safety management. The hierarchy is intended to have the following structure:

- Level 1 – key safety management requirements, including overarching objectives and principles, together with top-level guidance on accepted principles for complying with these requirements;

- Level 2 – more detailed guidance on accepted principles and practice, e.g. suggested contents for documents, checklists, etc.;

- Level 3 – information on specific techniques, e.g. fault trees, and technology specific issues, e.g. static analysis of safety critical software.

It is intended that level 1, i.e. DS 00-56, will be stable, but that levels 2 and 3 will evolve as necessary to reflect changes in technology and industrial best practice. At the time of writing, a task for the author is to "flesh out" this approach and to clarify the scope of the three levels. However, even at this stage, it is possible to outline the main idea in more detail.

Level 1 addresses the Limits and Objectives of the MoD safety policy, as set out above. Level 2 will amplify on the core requirements (Limit and Objectives) of DS 00-56, and will give more detailed guidance on MoD expectations, and provide mappings to existing civil standards (recall the "as civil as possible, and only as military as necessary" policy). Thus, for example, Level 2 will set out the expected contents of a hazard log, or requirements on safety case content (at different stages in the Acquisition process). It will also identify technology specific means of meeting the requirements of DS 00-56. This might include both a revised version of DS 00-55, and an identification of additional requirements over those of, say, DO178B (RTCA 1992) to meet the intent of the standard. Similarly, the role of IEC 61508 for dealing with the safety of E/E/PES will be dealt with at this level. It is likely that this Level will be quite contentious, and difficult to write. Plans for developing this level of material are contingent on the production of a better outline of the scope of this level, and a firmer draft of DS 00-56, i.e. Level 1.

Level 3 should be comparatively uncontentious, and simply refer to accepted standards and practices, e.g. the fault tree handbook (Roberts 1981). It is not anticipated that the MoD will have to produce much of level 3, however it may be that this is the right place for DS 00-58. Also, it is expected that it will be relatively easy to set out Level 3, once agreement has been reached on Levels 1 and 2.

Some consideration will also be given to the production of a "handbook" in support of DS 00-56. This may exist at Level 2, or Level 3 only, and it is not yet clear what the relationship of this material is to the three level structure outlined above – or indeed if it will turn out to be the realisation of Levels 2 and 3.

In the author's view, getting the shall/should balance right, i.e. facilitating the use of the ACARP principle, and ensuring that the right information is at the right level in the three tier structure are the central technical challenges of producing the new issue of DS 00-56.

As a final note, it is worth observing the proposed scope of the standard. It is concerned with avoiding or reducing:

- harmful effects to personnel, including MoD employees and civilians;
- material loss, including loss of defence equipment;
- environmental damage, including noise and pollution.

In other words, the scope has been expanded from DS 00-56 Issue 2 to incorporate environmental issues. This too poses challenges in developing the standard, although, fortunately, many of the techniques for safety management can be used for environmental issues as well.

4   Conclusions

Safety standards are very important. Although the principles of achieving and assuring safety are quite well-understood, there are considerable difficulties in implementing these principles in practice, e.g. deciding on the acceptable levels of risk. This is particularly difficult for defence systems where there are inevitable trade-offs between effectiveness and safety. Thus a key role for any UK defence standard is to articulate how acceptability of systems, equipments or services will be judged, and thereby to remove uncertainty from the Acquisition process.

The need for producing DS 00-56 Issue 3 has been established by a number of studies which have identified practical difficulties issues of principle with the current issue. Further, the Standards Breakthrough process requires a move to a form of standard which is less prescriptive and sets objectives rather than prescribing processes.

It is believed that the current draft is a good "step on the road" towards meeting the main objectives set out in section 3.1. However there are considerable challenges in meeting these objectives, and section 3.2 has highlighted some of the key issues. Also it is difficult to judge the role of the draft standard as a contractual document. It is hoped that further studies conducted on behalf of the SSRC will address this, to some degree, by soliciting industry views on the ability to work to the draft standard. In addition, if possible, some experimentation will be carried out to try to validate the principles of the draft standard, and to see how effective (e.g. clear, unambiguous, etc.) it might be as a contractual requirement.

Writing a standard to set requirements, without prescribing the means of meeting those requirements, is challenging. Perhaps surprisingly, there is little experience in writing objective-based standards. Some MoD standards, e.g. DS 00-40 (MoD 1999b), reflect this approach, thus have been used as a model where possible. However it is likely that some refinement of the current draft will be needed to meet the Standards Breakthrough philosophy.

In some respects moving to objective-based requirements is breaking new ground. However it is consistent with a wider move towards risk based regulation, e.g. by the CAA Safety Regulation Group (SRG) in respect of air traffic management.

It is very unlikely that DS 00-56 Issue 3, either now or when issued, will prove "the last word" on defence system safety. However it is hoped it will prove to be a useful step towards the production of a stable system safety standard which has widespread recognition and acceptance, at least in the defence community.

Finally, once DS 00-56 Issue 3 has been produced, it will give a context for the revision of lower level standards, especially DS 00-55.

5   Acknowledgements

Thanks are due to Peter Hardaker and Gareth Rowlands of the SSMO, respectively the Chairman and Secretary of the SSRC, for their support in undertaking this work. Colin O'Halloran of QinetiQ has provided technical input to the development of these ideas and has funded the work with the SVRC and MBDA, plus the drafting of DS 00-56 Issue 3. Peter Lindsay of the SVRC has influenced many of my ideas, particularly about the scope of the draft standard. David Smith of Frazer-Nash has led many of the studies mentioned above and provided a good insight into the views of the wider defence safety community. He has also contributed to the work described above through the DARP. Thanks are due to Tim Kelly for permission to use his (previously unpublished) ideas regarding the ACARP principle. Finally, I have had fruitful discussions with many other people over the years on a range of standards, including DS 00-56. They are too many to list here, but I should like to acknowledge their input and influence.

6 References

DoD (1993), MilStd 882C: System Safety Program Requirements.

DoD (1999), MilStd 882D: System Safety Program Requirements.

Frazer-Nash (2001) Report on the Survey of Defence Standards 00-55, 00-56 and 00-58, Available from the SSRC Secretary via ssrc@dpa.mod.uk

HIRTS DARP website; www.cs.york.ac.uk/hise/darp/.

International Electrotechnical Commission (1999), IEC61508: Functional Safety of Electrical/Electronic/Programmable Electronic Systems.

Lindsay P.A., McDermid J.A. (2002) Derivation of Safety Requirements for an Embedded Control System, in Proc. Systems Engineering, Test and Evaluation 2002, Sydney Australia.

MoD (1996a), Defence Standard 00-58, Issue 1, HAZOP Studies on Systems Containing Programmable Electronics.

MoD (1996b), Defence Standard 00-56, Issue 2, Safety Management Requirements for Defence Systems.

MoD (1997), Defence Standard 00-55, Issue 2, Requirements of Safety related Software in Defence Equipment.

MoD (1999a), Defence Standard 00-54, Requirements for Safety Related Hardware in Defence Equipment.

MoD (1999b) Defence Standard 00-40, Reliability and Maintainability, Part 1: Management Responsibilities and Requirements for Programmes and Plans.

MoD (2001), Joint Service Publication (JSP) 430: Ship Safety Management Code.

Roberts N.H., Vesely W.E, Haasl D.F., Goldberg F.F. (1981), NUREG 0492, Fault Tree Handbook, Nuclear Regulatory Commission.

McDermid J.A., Rowlands G., Wilson V.K. (2001), Notes from Puckrup Hall Workshop, 5th-6th November, Available from the SSRC Secretary via ssrc@dpa.mod.uk

RTCA (1992), Software Considerations In Airborne Systems and Equipment Certification, DO-178B/ED-12B

SAE (1996a), Aerospace Recommended Practice (ARP) 4754: Certification Considerations for Highly-Integrated or Complex Aircraft Systems.

SAE (1996b), Aerospace Recommended Practice (ARP) 4761: Guidelines and methods for conducting the safety assessment process on civil airborne systems and equipment.

AUTHOR INDEX

Armstrong, J. 63
Atkinson, D.C. 217
Burrett, G. 19
Cockram, T. 151
Daniels, D. 119
Donald, I. 43
Eastman, R. 137
Foley, S. 19
Gilchrist, I. 251
Hilton, A. 119
Howlett, R.F. 189
Johnson, S. 43
Jolliffe, G. 175
Kelly, T.P. 99
Lockwood, B. 151

Malcolm, B. 43
McDermid, J.A. 261
Moffat, N. 175
Myers, R. 119
Neil, M. 43
Nordland, O. 163
Paynter, S. 63
Popat, P. 137
Pygott, C.H. 201
Qiu Xie, C. 43
Shaw, R. 43
Sommerville, I. 3
Spriggs, J. 79
Strong, H.M. 217